History of Physics

The Springer book series *History of Physics* publishes scholarly yet widely accessible books on all aspects of the history of physics. These cover the history and evolution of ideas and techniques, pioneers and their contributions, institutional history, as well as the interactions between physics research and society. Also included in the scope of the series are key historical works that are published or translated for the first time, or republished with annotation and analysis.

As a whole, the series helps to demonstrate the key role of physics in shaping the modern world, as well as revealing the often meandering path that led to our current understanding of physics and the cosmos. It upholds the notion expressed by Gerald Holton that "science should treasure its history, that historical scholarship should treasure science, and that the full understanding of each is deficient without the other." The series welcomes equally works by historians of science and contributions from practicing physicists.

These books are aimed primarily at researchers and students in the sciences, history of science, and science studies; but they also provide stimulating reading for philosophers, sociologists and a broader public eager to discover how physics research – and the laws of physics themselves – came to be what they are today.

All publications in the series are peer reviewed. Titles are published as both print- and eBooks. Proposals for publication should be submitted to Dr. Angela Lahee (angela.lahee@springer.com) or one of the series editors.

Ricardo Lopes Coelho

What Is Energy?

An Answer Based on the Evolution of a Concept

Springer

Ricardo Lopes Coelho
Faculdade de Ciências
Universidade de Lisboa
Lisbon, Portugal

ISSN 2730-7549 ISSN 2730-7557 (electronic)
History of Physics
ISBN 978-3-031-51854-6 ISBN 978-3-031-51855-3 (eBook)
https://doi.org/10.1007/978-3-031-51855-3

This Springer imprint is published by the registered company Springer Nature Switzerland AG
The registered company address is: Gewerbestrasse 11, 6330 Cham, Switzerland

Paper in this product is recyclable.

Acknowledgements

I am grateful to the Instituto Bento da Rocha Cabral, where I have carried out my experimental work.

I would like to thank the colleagues who over the years have discussed and organized publications and talks on energy: António Passos Videira, Fabio Bevilacqua, Raffelo Pisano, Paulo Maurício, Falk Riess, Dietmar Höttecke, Panagiotis Kokkotas, Konstantinos Ravanis, Paulo Borges, Andreia Guerra, Marco Braga, Ricardo Karam, Mónica Batista, Manuel Bächthold, Peter Heering, Cibelle Silva, among others.

Thanks to Dr. Angela Lahee, Senior Editor at Springer-Verlag, for her suggestions and support.

Competing Interests

There are no Competing Interests

Concerning Permissions

Five figures, three of which have been modified, have appeared in my articles published in Springer journals (*Science & Education* 2009, 2014 and *Foundations of Science* 2021).

(Modified figures: 2.1, 2.2, 2.6; unchanged figures: 2.4, A2.1)

Contents

Chapter 1
Introduction

This chapter introduces the problem with the concept of energy and gives an overview of how it will be dealt with. In presenting the problem, physicists appear first, who from the end of the nineteenth century to the present day tell us that we have no answer to the question of what energy is. This is followed by disagreements between physicists, as some define or characterize energy and others criticize these theses. Finally, the authors will be considered who use the substance metaphor. Some of these know that energy is not a substance but talk about it as if it were, because they cannot find anything better. To find a solution to the problem, it would be necessary to determine what it consists of. To do this, it is important to capture its origin and how it developed. The introduction gives an overview of the chapters where this is done, as well as of the final chapter, which answers the question of the book.

1.1 The Problem

The History of Science teaches us that energy was discovered in the 1840s. Towards the end of the century the concept of energy became a problem. This problem has never been solved. This is the reason for the present book.

In the 1880's, Max Planck participated in a competition whose aim was to clarify the concept of energy. For this purpose, he wrote a book *The Principle of Conservation of Energy* (1887) in which, however, he pointed out a problem with the energy concept that he could not solve. This was the following. According to physics, the energy of an isolated system remains constant. Therefore, if energy is a substance, as it was understood, it must be somewhere in the system. It is, however, impossible to localise the energy in the system (Planck, 1921, p. 117). Hence, Planck considered that this concept of energy one day should be overcome (ibid. p. 118).

A few years later, the already well-known physicist Heinrich Hertz (1894) argued categorically: it is not logically permissible to take energy as a substance, since the

R. Lopes Coelho, *What Is Energy?*, History of Physics,
https://doi.org/10.1007/978-3-031-51855-3_1

properties of energy contradict the properties of a substance. He justifies. The quantity of a substance does not depend on the existence of other bodies. The potential energy of a body does depend on other bodies. Whereas the quantity of a substance is always a positive quantity, the potential energy of a system can be negative. Hence, energy should not be taken as a substance. A few years later, the French mathematical physicist Henri Poincaré came to the conclusion that "it is impossible to find a general definition" of energy. (Poincaré, 1901, p. 488) The reason was as follows. In each case of energy transformation, it was well known which energy forms were involved. Coming up with a concept of energy that encompassed all forms was, however, impossible.

In the twentieth century physics, new theories and forms of energy appeared. Despite this development, the Nobel laureate Richard Feynman points out 'It is important to realize that in physics today we have no knowledge of what energy *is*' (Feynman et al., 1963, Sect. 4–1).[1] Towards the end of the century, the first of Bergmann and Schaefer's eight volumes of *Experimental Physics* not only reinforces this idea, but also adds that a physicist is in the same situation as a layman when it comes to the question of what energy is (Bergmann & Schaefer, 1998, p. 616). According to Dransfeld et al. (2001, p. 109) 'we do not really know what energy is'. Other physicists have corroborated this same difficulty (Anderson, 2017; Çengel & Boles, 2002; Hecht, 2000).

Although there has been a problem with the concept of energy for decades and some physicists even say that we do not know what it is, other physicists define it. This definition, which tells us that energy is the capacity of doing work, has been criticised.[2] The point is, however, not the criticism of the definition but rather the fact that energy is defined by some physicists whereas others claim that we do not know what it is. One could expect that if energy is defined, we know what it is and, if we do not know, no definition can be given.

Some physicists do not use this definition of energy but instead the concept of energy which stems from the principle of the conservation of energy. This principle tells us that energy cannot be created or destroyed but only transformed. Trying to understand energy based on these properties, we can be led to think that energy is something that exists because we cannot destroy it and it would be meaningless to say that we cannot destroy what does not exist. As energy can only be transformed, we are led to think that energy appears in one form and later in another form. This reinforces the idea that energy is something that exists and manifests itself in different forms.

Is energy something that exists? Physicists diverge regarding this question. Some contemporary physicists take energy as a substance, as we shall see Sect. 5.2, whereas others are sure that energy is not a substance (Arons, 1999; Duit, 1987; Feynman et al., 1963; Hudson & Nelson, 1982). Contemporary physics, chemistry and biology textbooks express energy through the substance metaphor: energy can flow, be stored,

---

[1] In this book, the underlining in quotations is always from the original.

[2] This definition is addressed in Chap. 5 (Sect. 5.2.3) and the criticism of it by physics education experts in Appendix N.

transported, lost, can change its form, etc. (Lancor, 2014). Some experts in physics education propose to use the substance like metaphor in the teaching of energy (Brewe, 2011; Fortus et al., 2019; Scherr et al., 2012). Harrer claims that we have no alternative to the metaphorical expression of energy: "scientists were and are only able to investigate the multifaceted concept of energy by using models that employ conceptual metaphors" (Harrer, 2017, p. 454).

If the concept of energy is so difficult that we are doomed to express it only metaphorically, one wonders how it could have been discovered in the 1840s, when none of the discoverers (recognized as such by the History of Science) was a physicist. Did they discover something so important to science, as the principle of conservation of energy is but also so difficult that physicists even today do not know what it is? It seems to be so. Nevertheless, a short reflection leads us to another question. If we do not have a clear idea of what energy is, how can we say that what they discovered was energy? In fact, if what they discovered was energy, we only need to focus on what they discovered to know what energy is. As a matter of fact, research into the history of energy has been going on for decades[3] and we still have no answer to the question of what energy is. Under these circumstances, we first need to know what was discovered. This is the subject of the next chapter.

1.2 Overview of the Book's Development

In Chap. 2, four authors, who are usually taken as discoverers of energy by historians of science, will be addressed: Robert Mayer, James Joule, Ludvig Colding and Hermann von Helmholtz. Each of these authors defended a thesis contrary to the prevailing theory of heat, in which none of them was an expert. According to this theory, heat is a substance. This thesis had an experimental basis: experiments had shown that the quantity of heat was invariant in nature. Each of those four authors argued that the amount of heat does vary. The experimental basis of this argument lies above all in a finding by Mayer and Joule. These showed that the quantity of heat varies and this variation is correlated with the mechanical action involved. Indeed, they determined the number of mechanical units it takes to obtain one thermal unit. If this result is taken into account, the amount of heat varies and then heat cannot be a substance.

This conclusion created a new question: if heat it is not a substance, what is it then? Regarding this question, the four authors differed from each other. Mayer argued that heat is a form of *force*; Joule said that heat is *motion*; Colding argued that it is a *force*

[3] Planck (1921 [1887]), Mach (1896), Helm (1898), Haas (1909), Kuhn (1959), Theobald (1966), Breger (1982), Schirra (1989), Smith (1998), Guedj (2000), Müller (2007), Coelho (2009), Coopersmith (2015), and Hanlon (2020). On Mayer: Weyrauch (1890), Riehl (1900), Hell (1914), Timerding (1925), Mittasch (1940), Lindsay (1973), Heimann (1976), Smith (1978), Caneva (1993); On Joule: Fox (1969), Forrester (1975), Cardwell (1989); on Colding: Dahl (1963), Kragh (2009); on Helmholtz: Elkana (1974), Heimann (1974), Bevilacqua (1983, 1993), Ordónez (1996), Caneva (2021). See also Appendix M.

of nature; Helmholtz defended that heat is *motion*. The idea that heat is motion still agrees with the theory of heat of the time in the extent to which heat as motion was taken as an alternative to heat as a substance. (Indeed, some physicists and chemists argued that heat was a substance,[4] others that it was motion,[5] and still others were in doubt about these two theses.[6]) Heat as a form of force was new, it was strange to the heat theory of the time. According to whether they adopted the concept of heat as motion or heat as force, the same phenomena were seen as 'conversion' or 'transformation' respectively. If heat is a form of force and if, for instance, heat is produced by a mechanical action, this phenomenon was considered a *transformation* of one force into the other. If heat is motion, the same phenomenon was considered as a *conversion* of one kind of motion into another. For those who take heat as motion, the kind of motion was a further question. Thus, in the 1840s, different theories for the same phenomena appeared.

Chapter 3 begins with the question that arises among experts in the theory of heat in the late 1840s. On the one hand, there was an experimental finding: the value of the mechanical units necessary to obtain one thermal unit, what was called the mechanical equivalent of heat. According to this experimental research, heat can be produced by mechanical action or the other way around. In the first case, the quantity of heat increases and in the second, it decreases. Since the quantity of heat can vary, it cannot be a substance. On the other hand, there was a theory of heat based on the concept of heat as a substance, Carnot's theory of heat. Abandoning this theory because of those experimental results was out of the question, according to William Thomson (later Lord Kelvin) in 1849, because Carnot's theory was complete and to abandon it would create difficulties.

A new input comes then from Clausius (1850). He solves the incompatibility between Carnot's theory and Joule's experiments and presented the two laws of thermodynamics. One law assumes Joule's experiments and the other, Carnot's contribution slightly modified. In 1851, Thomson accepts the thesis that heat is motion and creates a new concept: the *mechanical activity* of a body. 'Mechanical activity' comprises what a body can do from a strictly mechanical point of view and what it can do by virtue of its heat. (Heat could be subsumed by 'mechanical activity' since heat had become a kind of motion.) This concept of mechanical activity is called, by Thomson, the *mechanical energy* of a body. This is how 'energy', an erudite term for activity, was introduced into the heat theory.

Since mechanics was divided into statics and dynamics, Thomson (1852) systematizes the reserves of *mechanical energy* available to mankind into reserves of the static kind and reserves of the dynamic kind. In 1853, Rankine adopted Thomson's definition of mechanical energy but changes his systematization of the reserves. Instead of 'dynamic', Rankine uses 'actual' and what was 'static' becomes 'potential'. This is not only a change of name, because whereas the static ended with the

[4] Karsten (1790), Haldat 1807, Berthollet (1809), Suckow (1813), Carnot (1824), Colladon and Sturm (1828), Clapeyron (1834).

[5] Rumford (1798), Davy (1799), Young (1807), Mayer (1820).

[6] Laplace and Lavoisier (1780), Black (1803), Kämtz (1839) (see Appendix A).

beginning of motion, potential energy varies continuously with the dynamic. Thus, it could be said that if one of them increases, the other decreases in the same proportion. This couple *actual* and *potential*, which has its origin in Aristotle's philosophy, is the origin of what we call kinetic and potential energy.

Several of the concepts considered above were included in textbooks. In Maxwell's *Theory of Heat*, Rankine's definition of energy, Thomson's definition of mechanical energy of a body, Rankine's forms of energy (kinetic and potential) and Mayer's forms of energy (his term 'force' was translated by 'energy') appear together but without referring to the authors' name. Thus, what were different points of view on energy appeared as characteristic of energy. Mayer's forms of energy and Rankine's were not seen as different theses about energy but simply forms of energy, which means then that there were two kinds of forms of energy. Understanding energy with all these eclectic associations became difficult.

Chapter 4 is devoted to a new phase of the concept: energy becomes a substance. In 1873, Maxwell published his *Treatise on Electricity and Magnetism.* This book will play a role in the history of the concept of energy. Accepting Maxwell's concept of electromagnetic field, Poynting arrives at the idea that energy moves through space. This thesis stems from an interpretation of the algorithm used by him to account for electromagnetic phenomena (Poynting, 1884). He himself points out, however, that experimental work, which could corroborate or deny the thesis, was still lacking. Even though without experimental support, Lodge (1885) defends the thesis that energy moves in space and transfers from one body to another or between bodies and the ether. This conceptual framework is not only applied to electromagnetic phenomena but also to the falling of a stone or the motion of a pendulum. A pendulum oscillates, Lodge explains, because energy comes from the aether to the bob of the pendulum and back from the bob to the aether, depending on whether the bob is going up or down. The concept of aether lost its significance in the twentieth century physics. Lodge's theory appears in contemporary textbooks reduced to the idea that energy is something that is possessed by bodies and transferred between them.

Another reification of energy comes from Ostwald, who was the Chemistry Nobel Prize winner in 1909. He takes up Mayer's theory but unlike him he substantializes energy. Energy becomes the real thing. Ostwald explains, it is the real thing because it is the cause of events and their content. Matter and spirit are subsumed by this concept of energy: there is no matter and no spirit but only energy. The generalization of energy to the various fields of human activity, including ethics, is rooted here.

When in the light of this historical background, one looks at contemporary textbooks on general physics (bachelor level), what will be done in Chap. 5, one realizes that contemporary concepts of energy and principles of conservation come from different theses on the subject that appeared in the nineteenth century. This in itself is no objection. On the contrary, it would be nonsense to waste the work done in the nineteenth century. The point at issue is only that the theses are not compatible with each other. Sometimes they are even contradictory. In order to clarify these divergences, contemporary theses are linked to their origin in that chapter. This enables us to verify whether a given thesis was well-founded when it was presented for the first time or in the course of time.

The historical texts that are addressed in the following chapters serve to capture the origin of the problem with the concept of energy and how it developed. The analysis of these texts is presented in such a way that the reader can follow it step by step. The final chapter presents the result of this study. This shows where the *problem* presented above lies. This is also the solution we need, because on this basis the divergences between physicists are then explained and conceptual difficulties with energy are overcome. What can be done from this point on is an open question.

Topics from the history of energy that are not in the problem line but are useful for understanding the texts under analysis are dealt with in the appendices. The same goes for consequences of the problem of the concept of energy in education, where experts have recorded student misconceptions and criticized textbooks. Specific questions concerning the historiography of energy will also be addressed in the appendices.

References

Anderson, G. M. (2017). *Thermodynamics of natural systems: Theory and applications in geochemistry and environmental science* (3rd ed.). Cambridge University Press.

Arons, A. B. (1999). Development of energy concepts in introductory physics courses. *American Journal of Physics, 67*, 1063–1067.

Bergmann, L., & Schaefer, C. (1998). *Lehrbuch der Experimentalphysik* (11th ed., Vol. I). de Gruyter.

Berthollet, C. L. (1809). Notes sur divers objects. *Mémoires de Physique et de Chimie de la Société d'Arcueil*. Tome II, 441–448. Johnson.

Bevilacqua, F. (1993). Helmholtz' Ueber die Erhaltung der Kraft. In D. Cahan (Ed.), *Hermann von Helmholtz and the foundations of the nineteenth-century science* (pp. 291–333). University of California Press.

Bevilacqua, F. (1983) *The principle of conservation of energy and the history of classical electromagnetic theory*. La Goliardica Pavese.

Black, J. (1803) *Lectures on the elements of chemistry, delivered in the University of Edinburgh* (Vol. 1). In John Robison (ed.) Creech.

Breger, H. (1982). *Die Natur als arbeitende Maschine: Zur Entstehung des Energiebegriffs in der Physik 1840–1850*. Campus Verlag.

Brewe, E. (2011). Energy as a substance like quantity that flows: Theoretical considerations and pedagogical consequences. *Physical Review Special Topics Physics Education Research, 7*(020106), 1–14.

Caneva, K. L. (1993). *Robert Mayer and the conservation of energy*. Princeton University Press.

Caneva, K. L. (2021). *Helmholtz and the conservation of energy: Contexts of creation and reception*. MIT Press.

Cardwell, D. S. L. (1989). *James Joule. A biography*. Manchester University Press.

Carnot, S. (1824). *Réflexions sur la puissance motrice du feu*. Bachelier. (Rep. Paris: Éditions J. Gabay, 1990).

Çengel, Y., & Boles, M. (2002). *Thermodynamics*. Mc Graw Hill.

Clapeyron, B. (1834). Mémoire sur la puissance motrice de la chaleur. *Journal de l'École Polytechnique, 14*, 153–190.

Coelho, R. L. (2009). On the concept of energy: How understanding its history can improve physics teaching. *Science & Education, 18*, 961–983.

Colladon, J.-D., & Sturm, C. F. (1828). Ueber die Zusammendrückbarkeit der Flüssigkeiten. *Annalen Der Physik, 88*, 161–197.

Coopersmith, J. (2015). *Energy, the subtle concept: The discovery of Feynman's blocks from Leibniz to Einstein* (Revised). Oxford University Press.

Dahl, P. F. (1963). Colding and the conservation of energy. *Centaurus, 8*, 174–188.

Davy, H. (1799) An essay on heat, light, and combinations of light. In: J. Davy (ed.) *The Collected Works of Humphry Davy* (vol. 2, pp. 2–86). Smith, Elder and Co.

Dransfeld, K., Kienle, P., & Kalvius, G. M. (2001). *Physik I: Mechanik und Wärme* (9th ed.). Oldenbourg.

Duit, R. (1987). Should energy be illustrated as something quasi-material? *International Journal of Science Education, 9*, 139–145.

Elkana, Y. (1974). *Discovery of the conservation of energy*. Hutchinson.

Feynman, R. P., Leighton, R. B. & Sands, M. (1963). *The Feynman lectures on physics* (Vol. 1). Addison-Wesley.

Forrester, J. (1975). Chemistry and the conservation of energy: The work of James Prescott Joule. *Studies in History and Philosophy of Science, 6*, 273–313.

Fortus, D., Kubsch, M., Bielik, T., Krajcik, J., Lehavi, Y., Neumann, K., Nordine, J., Opitz, S., & Touitou, I. (2019). Systems, transfer, and fields: Evaluating a new approach to energy instruction. *Journal of Research in Science Teaching, 56*, 1341–1361.

Fox, R. (1969). James Prescott Joule (1818–1889). In John North (Ed.), *Mid-nineteenth-century Scentists* (pp. 72–103). Pergamon Press.

Guedj, M. (2000). *L'émergence du principe de conservation de l'énergie et la construction de la thermodynamique*. (Diss.) Paris.

Haas, A. (1909). *Die Entwicklungsgeschichte des Satzes von der Erhaltung der Kraft*. Hölder.

Haldat. (1807). Recherches sur la chaleur produite par le frottement. *Journal De Physique, De Chimie Et D'histoire Naturelle, 65*, 213–222.

Hanlon, R. (2020). *Block by block: The historical and theoretical foundations of thermodynamics*. Oxford University Press.

Harrer, B. W. (2017). On the origin of energy: Metaphors and manifestations as resources for conceptualizing and measuring the invisible, imponderable. *American Journal of Physics, 85*, 454–460.

Hecht, E. (2000). *Physics: Calculus* (2nd ed., Vol. 1). Brooks/Cole.

Heimann, H. (1974). Helmholtz and Kant: The metaphysical Foundations of *Ueber die Erhaltung der Kraft*. *Studies in History and Philosophy of Science, 5*, 205–238.

Heimann, H. (1976). Mayer's concept of force: The axis of a new science of physics. *Historical Studies in the Physical Sciences, 7*, 277–296.

Hell, B. (1914). Robert Mayer. *Kantstudien, 19*, 221–248.

Helm, G. (1898). *Die Energetik nach der geschichtlichen Entwicklung*. Veit & C.

Hertz, H. (1894). *Die Prinzipien der Mechanik*. J. A. Barth.

Hudson, A., & Nelson, R. (1982). *University Physics*. H. B. Jovanovich.

Kämtz, L. F. (1839). *Lehrbuch der Experimentalphysik*. Gebauer

Karsten, W. J. G. (1790). *Anfangsgründe der Naturlehre* (F. Gren, Ed. 2nd ed.). Halle: Rengersche Buchhandlung.

Kragh, H. (2009). Conservation and controversy: Ludvig Colding and the imperishability of "forces". *RePoSS: Research Publications on Science Studies* 4. Department of Science Studies, University of Aarhus.

Kuhn, T. S. (1959). Energy conservation as an example of simultaneous discovery. In M. Clagget (Ed.), *Critical problems in the history of science* (pp. 321–356). Wisconsin University Press.

Lancor, R. (2014). Using metaphor theory to examine conceptions of energy in biology, chemistry, and physics. *Science and Education, 23*, 1245–1267.

Laplace, P. S., Lavoisier, A. (1780) Mémoire sur la chaleur. In *Oeuvres Complètes de Lavoisier.* Vol. 10, pp. 149–200. Gauthier-Villars.

Lindsay, R. (1973). *Julius Robert Mayer*. Pergamon Press.

Mach, E. (1896). *Principien der Wärmelehre. Historisch-kritisch entwickelt*. J. A. Barth.

Mayer, J. T. (1820). *Anfangsgründe der Naturlehre* (4th ed.). Dieterich Buchhandlung.

Mittasch, A. (1940). *Julius Robert Mayers Kausalbegriff*. Springer.
Müller, I. (2007). *A history of thermodynamics: The doctrine of energy and entropy*. Springer.
Ordónez, J. (1996). The story of a non-discovery: Helmholtz and the conservation of energy. In G. Munévar (Ed.), *Spanish Studies in the Philosophy of Science* (pp. 1–18). Kluwer.
Planck, M. (1921 [1887]). *Das Prinzip der Erhaltung der Energie*. 4th ed. Teubner.
Poincaré, H. (1901). Sur les Principes de la Mécanique. In: *Ier Congrès International de Philosophie*, Tome 3. Paris, 1900, pp. 457–494 (Reprint Nendel, Liechtenstein: Kraus Reprint Limited, 1968).
Riehl, A. (1900). Robert Mayers Entdeckung und Beweis des Energieprincipes. In C. Sigwart, B. Erdmann (eds.) *Philosophische Abhandlungen*. Tubingen, Freiburg i. B., Leipzig: J.C.B. Mohr.
Rumford, B. C. (1798). An inquiry concerning the source of the heat which is excited by friction. *Philosophical Transactions of the Royal Society of London 88*, 80–102.
Scherr, R. E., Close, H. G., McKagan, S. B., & Vokos, S. (2012). Representing energy. I. Representing a substance ontology for energy. *Physical Review Special Topics Physics Education Research, 8*, 020114.
Schirra, N. (1989). *Entwicklung des Energiebegriffs und seines Erhaltunskonzepts*. Dissertation. Justus-Liebig-Universität.
Smith, C. (1978). A new chart for British natural philosophy: The development of energy physics in the nineteenth century. *History of Science, 16*, 231–279.
Smith, C. (1998). *The science of energy: A cultural history of energy physics in Victorian Britain*. The Athlone Press.
Suckow, G. A. (1813). *Anfangsgründe der Physik und Chemie nach den neuesten Entdeckungen*. Statische Buchhandlung.
Theobald, D. (1966). *The concept of energy*. Spon.
Timerding, H. (1925). *Robert Mayer und die Entdeckung des Energiegesetzes*. Deuticke.
Weyrauch, J. (1890). *Robert Mayer, der Entdecker des Princips von der Erhaltung der Energie*. K. Wittwer.
Young, T. (1807). *Thomas Young's Lectures on Natural Philosophy and the Mathematical Arts*, I. (Rep. Bristol: Thoemmes, 2002)

Chapter 2
What Was Discovered in the 1840s?

This chapter analyzes the texts of the authors considered to be the discoverers of energy: Robert Mayer, James Joule, Ludvig Colding and Hermann von Helmholtz. Although they have in common that they defend a thesis on heat that was contrary to the science of the time, they differ in their interpretation of the experiments, namely on the nature of heat, and in the theories they developed. This diversity was mitigated when they began to appear as having discovered the same thing. The present analysis aims to provide a sharper idea of the authors by focusing on the texts of each of them and making a systematic distinction between experiments and interpretations. After that, some of their theses will be compared with each other and with the current principle of conservation of energy.

2.1 Robert Mayer

Mayer was a physician in clinical activity. The idea of conservation of energy comes to him from this practice. He tells us that on a boat trip to Indonesia, where he was the doctor on duty, he performed phlebotomies on all the crew members, because they all had a lung infection. He was amazed at the clear color of the venous blood.[1] The venous blood was much clearer than he knew from his clinical experience in Central Europe. He then concluded that there was a relationship between blood color and ambient temperature. The venous blood was darker in Central Europe

[1] "Während einer hunderttägigen Seereise war bei der aus 28 Köpfen bestehenden Equipage kein erheblicher Krankheitsfall vorgekommen; wenige Tage aber nach unserer Ankunft auf der Rhede von Batavia verbreitete sich epidemisch eine acute (katarrhalisch-entzündliche) Affection der Lungen. Bei den reichlichen Aderlässen, welche ich machte, *hatte aus der Armvene gelassene Blut eine ungemeine Röthe, so, dass ich der Farbe nach glauben konnte, eine Arterie getroffen zu haben.* [...] Bei einer reichlichen Aderlässe, welche ich zwei Monate nach unserer Ankunft in Java an einem kräftigen, von einer Leberentzündung befallenen Matrosen anstellte, fand ich eine normale schwarze Farbe des Blutes" (Mayer, 1845, pp. 84–5) (Appendix C).

R. Lopes Coelho, *What Is Energy?*, History of Physics,
https://doi.org/10.1007/978-3-031-51855-3_2

because maintaining body temperature required more oxygen consumption than in a tropical zone.[2] Thus, emerged the idea that became the germ of energy: for something to happen (keep the body temperature in a colder area), something has to be done (consume more oxygen). 'Nothing comes from nothing,' 'nothing becomes nothing,' 'cause equals effect'. These are the ways he expresses that idea in his published texts.

At that time, the word 'energy' existed in the lexicon but not as a scientific term. It meant 'activity'.[3] Mayer did not use the term energy but rather 'force'. So, he tells us about the *indestructibility* and *transformability* of force. This characterization of force is transferred to energy shortly after the appearance of this term in the heat theory, which occurred in 1851.

2.1.1 1842: Heat, Motion and the Equivalent

In 1842, Mayer publishes his first article "Observations on the forces of inorganic nature" in the German journal, *Annals of Chemistry and Pharmacy.* This paper addresses two questions: what is to be understood by forces; and how they relate to each other.[4] The answer to the first is "forces are causes." This statement is not justified. It plays, however, an important role. It serves to apply to forces a classical proposition about causes, "*causa aequat effectum*" (cause equals to effect).[5] The properties that he draws for causes on the basis of this proposition will then hold true for forces. (These properties will be adopted for energy later.)

---

[2] "Aus den bisher betrachteten Gesetzen folgt mit Nothwendigkeit, dass der *Temperatur*unterschied zwischen der Eigenwärme des Organismus und der Wärme des umgebenden Mediums in einer Grössenbeziehung mit dem *Farben*unterschiede beider Blutarten, des Arterien- und des Venenblutes stehen müsse. Je grösser dieser Temperaturunterschied, oder die Kraftproduktion, um so grösser muss auch der Farbenunterschied, und je kleiner der Unterschied der Temperatur, um so kleiner auch der der Farbe seyn. Dieser Farbenunterschied ist ein Ausdruck für die Grösse des Sauerstoffverbrauches, oder für die Stärke des Verbrennungsprocesses im Organismus" (ibid. pp. 85–6).

[3] Energy etymologically means activity. The term was used in the eighteenth century. The psychological sense, for example, was lexicalized in 1798 (Delon, 1988, p. 45). In 1807, Thomas Young proposes to use 'energy' to define the quantity, which was called living force, i.e. mass times the square of velocity (Young, 1807 pp. 78–9). Throughout the first half of the nineteenth century, the term energy appears in German, French and English scientific texts with the sense of activity: Seebeck uses it in his lessons (Seebeck, 1822–23, p. 265); Ampère speaks of energy around the same time (Ampère, 1822, p. 60); Faraday uses it in 1832 (Faraday, 1832, p. 133, § 34). Mayer uses the term in various contexts: he speaks of energy of mechanical effects, (Mayer, 1845, p. 28) energy of oxidation processes, (ibid. p. 79); the energy of heat or heat radiation. (ibid. 1848, pp. 23–5) Thomson also uses the term in a footnote in the 1849 article on Carnot's theory, in which he asks the question: since nothing can be lost in natural operations, *no energy can be destroyed*, what about heat that produces no mechanical effect? (Thomson, 1849, p. 545).

[4] "Der Zweck folgender Zeile ist, die Beantwortung der Frage zu versuchen, was wir unter "Kräften" zu verstehen haben, und wie sich solche untereinander verhalten" (Mayer, 1842, p. 233).

[5] "Kräfte sind Ursachen, mithin findet auf dieselbe volle Anwendung der Grundsatz: *causa aequat effectum*" (ibid. p. 233).

If the cause is equal to the effect and this effect is, in turn, the cause of another effect, then the first cause is equal to this second effect ($C = E$ and $E = E_2$, so $C = E_2$). This second effect can, in turn, become the cause of another effect and so on. Therefore, $C = E = E_2 = E_3 = \ldots$. In this sequence, no element can be greater or less than any other. From this Mayer derives a property of causes, that causes are indestructible.[6]

So far, we have talked about cause and effect in general. Let us now focus on a relationship in which we have a *given cause* and a *given effect*. Here we realize that the cause differs from the effect. We have already admitted, however, that the cause is equal to the effect. Hence, we have for our special case of given cause C and given effect E, that $C = E$. We know, however, that C and E differ from each other. Mayer makes the two sides compatible as follows: C transforms into E. From here comes another property of causes: causes are transformable.

Now we can return to the initial proposition, "forces are causes." Since forces are causes, the properties of causes apply to forces. Causes are indestructible and transformable. Therefore, forces are indestructible and transformable. The following then follows. If in a given phenomenon, we have a force F_1, which corresponds to the cause, and a force F_2, which corresponds to the effect, then we can write $F_1 = F_2$, by virtue of indestructibility, and say that F_1 is transformed into F_2, by virtue of the transformability of forces.

So far, we have had a theory. How does it apply to phenomena? To deal with phenomena, Mayer goes back to the cause-effect relationship. First, he seeks to determine whether in a given phenomenon there is a causal relationship. If so, if A is the cause of B, then A and B are forces. Then, it holds for A and B what we have seen for forces. Let us move on to the phenomena addressed in the paper.

Motion

Mayer begins by telling us, how to understand cause and effect in phenomena of lifting and falling of bodies. The lifting and falling of a body is understood as follows: there is a cause that lifts the body, therefore a force; the body lifted is also a force, called the falling force; its effect, the downward motion, is also a force.[7] Therefore, Mayer continues, they transform into each other.[8]

[6] "Hat die Ursache *c* die Wirkung *e*, so ist $c = e$; ist *e* wieder die Ursache einer andern Wirkung *f*, so ist $e = f$, u.s.f. $c = e = f \ldots = c$. In einer Kette von Ursachen und Wirkungen kann, wie aus der Natur einer Gleichung erhellt, nie ein Glied oder ein Theil eines Gliedes zu Null werden. Diese erste Eigenschaft aller Ursachen nennen wir ihre *Unzerstörlichkeit*" (ibid. p. 233).

[7] "Eine Ursache, welche die Hebung einer Last bewirkt, ist eine Kraft; ihre Wirkung, *die gehobene Last*, ist also ebenfalls *eine Kraft*; [...] da diese Kraft den Fall der Körper bewirkt, so nennen wir sie *Fallkraft*" (ibid. p. 235).

[8] "Fallkraft und Fall, und allgemeiner noch Fallkraft und Bewegung sind Kräfte, die sich verhalten wie Ursache und Wirkung, Kräfte, die in einander übergehen, zwei verschiedene Erscheinungsformen eines und desselben Objectes" (ibid. p. 235).

Quantitative

The falling force is given by the product of the mass of the body and by the height it is at (written with Mayer's symbols, *md*).[9] Due to the relationship between height and falling velocity established in the seventeenth century, the height is equal to the square of the velocity (with Mayer's symbols $d = c^2$).[10] Now, since the force of fall equals mass times height (*md*) and height equals the square of velocity ($d = c^2$), it follows that the force of fall equals mass times the square of velocity ($md = mc^2$) (Mayer, 1842, p. 236).

Since, according to Mayer's theory, force is indestructible, it follows that the quantity 'mass times the square of velocity' is conserved. The conservation of this magnitude, called 'living forces', was defended by Leibniz (1686) (Appendix B). Mayer argues then that the principle of conservation of living forces is based on his law of the indestructibility of causes.[11]

Connection with the science of that time

In the Mechanics of that time, weight was said to be a force and the cause of falling. Mayer argues against this thesis. The lifting of the weight is no less necessary to the fall than weight. Therefore, weight cannot be seen as the cause of falling.[12] From the point of view of his own theory he argues: designating weight by force contradicts the characteristics of force—being indestructible and transformable—because weight does not decrease with the fall.[13] Thus, he claims that weight is a property[14] and that the force of falling consists in the spatial difference of ponderable objects.[15]

---

[9] "Die Größe der Fallkraft v steht [...] mit der Größe der Masse m und mit der ihrer Erhebung d, in geradem Verhältnisse; $v = md$" (ibid. p. 236).

[10] "Geht die Erhebung $d = 1$ der Masse m in Bewegung dieser Masse von der Endgeschwindigkeit $c = 1$ über, so wird auch $v = mc$; aus den bekannten zwischen d und c stattfindenden Relationen ergiebt sich aber für andere Werthe von d oder c, mc^2 als das Maß der Kraft v" (ibid. p. 236).

[11] "also $v = md = mc^2$; das Gesetz der Erhaltung lebendiger Kräfte finden wir in dem allgemeinen Gesetze der Unzerstörbarkeit der Ursachen begründet" (ibid. p. 236).

[12] "Um daß ein Körper fallen könne, dazu ist seine Erhebung nicht minder nothwendig, als seine Schwere, man darf daher auch letzterer allein den Fall der Körper nicht zuschreiben" (ibid. p. 236).

[13] "gerade das, was jeder Kraft wesentlich zukommen muß, die *Vereinigung* von Unzerstörlichkeit und Wandelbarkeit, geht jedweder Eigenschaft ab [...] Heißt man die Schwere eine Kraft, so denkt man sich damit eine Ursache, welche, ohne selbst abzunehmen, Wirkung hervorbringt, hegt damit also unrichtige Vorstellungen über den ursächlichen Zusammenhang der Dinge" (ibid. pp. 235–6).

[14] "Indem man die Schwere als Ursache des Falls betrachtet, spricht man von einer Schwerkraft und verwirrt so die Begriffe von Kraft und Eigenschaft [...] zwischen einer Eigenschaft und einer Kraft, zwischen Schwere und Bewegung läßt sich deßhalb auch nicht die für ein richtig gedachtes Causalverhältniß nothwendige Gleichung aufstellen" (ibid. pp. 235–6).

[15] "*räumliche Differenz ponderabler Objecte ist eine Kraft*; da diese Kraft den Fall der Körper bewirkt, so nennen wir sie *Fallkraft*" (ibid. p. 235).

Heat and motion

Mayer approaches phenomena involving heat and motion in the same way as the previous ones. First, he tries to make sure, whether a cause-effect relationship can be established between both motion and heat.[16] He concludes from his observation that such a relationship exists. Then heat and motion are taken as forces.

Among his observations that led him to that idea, one was particularly important. He reports an experiment that he had carried out: a vigorous agitation of the water in a test tube raised its temperature by 12° centigrade. This experiment about which no details are given, was decisive for Mayer's thesis. One sees that motion disappears and heat appears. It can be said that the motion performed had an effect. Then motion can be seen as the cause of the heat. In light of this experiment and "many" other cases (he refers to everyday observations), Mayer argues that he prefers to admit that heat comes from motion, rather than admit that there is a cause without effect or an effect without cause.[17] Steam engines provide Mayer with an example of the reverse causation, from heat into motion.[18] Having admitted the cause-effect relationship for motion and heat, motion and heat are then forces. Mayer then moves on to the numerical determination of this relationship, which has the following experimental basis.

Quantitative

It was known at that time that if a gas is heated and the piston of the container in which the gas is, is movable, the volume increases but the pressure remains the same. On the contrary, if the piston does not move, the pressure increases but the volume does not change. It was also known that the heat required to raise the temperature of a certain amount of atmospheric air when the piston is moveable was greater than the heat required to raise the same amount of air from the same temperature when the piston is fixed (Fig. 2.1). Now, since in the first case, there is movement (the gas moves the piston), and in the second, there is not, but in the first case more heat is required than in the second, Mayer relates this excess of heat to the movement made.

Using the experimental values available at that time, Mayer was able to determine which quantity of heat corresponds to that motion. Now, he made another step. If to a quantity of heat corresponds a certain mechanical value, then to the thermal unit corresponds x. Thus, he arrived at the *mechanical equivalent of heat*. This is expressed

[16] "Um uns über dieses Verhältniß zu vergewissern, müssen wir die Frage erörtern, hat nicht in den zahllosen Fällen, in denen unter Aufwand von Bewegung Wärme zum Vorschein kommt, die Bewegung eine andere Wirkung als die Wärmeproduktion und die Wärme eine andere Ursache als die Bewegung?" (ibid. p. 237).

[17] "Ist es nun ausgemacht, daß für die verschwindende Bewegung in vielen Fällen (exceptio confirmat regulam) keine andere Wirkung gefunden werden kann, als die Wärme, für die entstandene Wärme keine andere Ursache als die Bewegung, so ziehen wir die Annahme, Wärme entsteht aus Bewegung, der Annahme einer Ursache ohne Wirkung und einer Wirkung ohne Ursache vor" (ibid. p. 238).

[18] "umgekehrt dienen wieder die Dampfmaschinen zur Zerlegung der Wärme in Bewegung oder Lasterhebung. Die Locomotive mit ihrem Convoi ist einem Destillirapparate zu vergleichen; die unter dem Kessel angebrachte Wärme geht in Bewegung über" (ibid. p. 239).

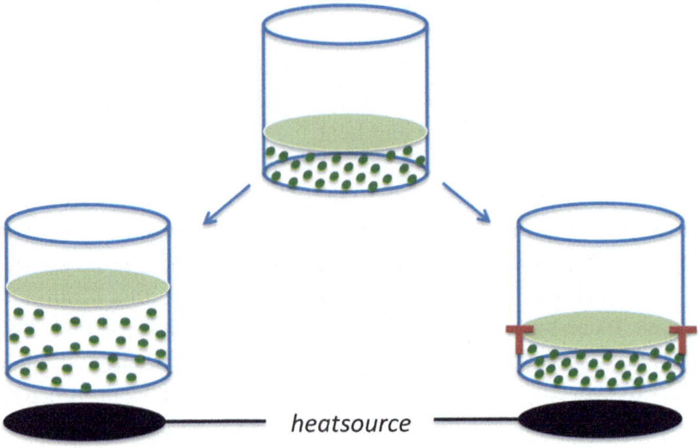

Fig. 2.1 The center container represents the initial situation of the gas in the container. Placed on the heat source, the gas heats up and expands (on the left). In this case, the pressure is constant. If the plunger is fixed (right), the gas heats up without changing its volume

as follows: heating a given quantity of water by 1° centigrade, more exactly from 0 to 1 °C, corresponds to the falling of a body of equal mass from a height of 365 m.[19] (In current units, it is 1 cal = 3.58 J.)

Conclusion

In Mayer's dealing with phenomena, we have two steps:

- Determine whether there is a cause-effect relationship;
- Establish the quantitative relationship between the magnitudes involved. The mechanical equivalent of heat is a consequence of this relationship.

The attribution to a phenomenon of the cause-effect relationship is based on observation. Mayer makes sure that he can establish a relationship between observable elements of the phenomenon at stake. A body at some point, if dropped, falls. This is the reason for attributing the cause-effect relationship to this phenomenon. In the other case, he stirred water and the water heated up. Thus, the motion is cause and the heat, effect. The steam engine received heat from the furnace and produced a pendulum motion at the top. So, heat was the cause and motion, the effect. When the cause-effect relation applies to a phenomenon, then it follows that 'cause = effect' applies. Now, when he deals with phenomena, he had already introduced

[19] "Unter Anwendung der aufgestellten Sätze auf die Wärme- und Volumensverhältnisse der Gasarten findet man die Senkung einer ein Gas comprimirenden Quecksilbersäule gleich der durch die Compression entbundenen Wärmemenge und es ergiebt sich hieraus,—den Verhältniß-exponenten der Capacitäten der atmosphärischen Luft unter gleichem Drucke und unter gleichem Volumen = 1,421 gesetzt, daß dem Herabsinken eines Gewichtstheiles von einer Höhe circa 365^{m}, die Erwärmung eines gleichen Gewichttheiles Wasser von 0° auf 1° entspreche" (ibid. p. 240).

the concept of force, so ‘force (cause) = force (effect)’ applies. This justifies then Mayer’s equalization of the quantities representing force-cause and force-effect:

- mass times height (force-cause) with the product of mass by the square of velocity (force-effect);
- the difference between the values of the heats at constant pressure and constant volume (force-cause) with weight times height (force-effect).

2.1.1.1 Concerning the Principle of Conservation of Energy

Determinative for the acceptance of the principle of the conservation of energy was the mechanical equivalent of heat as we shall see. Mayer determined it for the first time in history in this article.

The usual formulation of the principle of the conservation of energy says: energy can neither be created nor destroyed, but only transformed. If energy cannot be destroyed, then it is indestructible; if it is transformed, then it is transformable. Here we have the characteristics of Mayer’s force: indestructibility and transformability.

‘Indestructibility’ tells us that the quantity of force does not change. With regard to transformability, Mayer felt the need to clarify it. Motion turning into heat tells us nothing about the nature of heat. He does not consider for example that heat is motion, what it would mean to know something about the nature of heat. On the contrary, he even intends to conclude the opposite, that is, that motion disappears to give way to heat,[20] which corresponds to what is observed.

In sum, the concept ‘force is indestructible and transformable’ was created by Mayer but it was not designed to convey knowledge about the phenomena at stake that is beyond the observable or measurable data. The nature of heat, which refers to a deep knowledge of phenomena, is not at issue.

2.1.2 *1845: Forms of Force and the Equivalent*

In 1845, Mayer published a book at his own expense, *The Organic Movement in Connection with Metabolism.* Forces are now systematized into five forms: falling force, motion, heat (which already appeared in the 42 article), magnetism and electricity, separation and chemical bonding. The task of physics would then be to prove the “metamorphoses” of the 5 forms through 25 types of experiments.[21] Completely

[20] “So wenig indessen aus dem zwischen Fallkraft und Bewegung bestehenden Zusammenhange geschlossen werden kann: das Wesen der Fallkraft sey Bewegung, so wenig gilt dieser Schluß für die Wärme. Wir möchten vielmehr das Gegentheil folgern, daß um zu Wärme werden zu können, die Bewegung […] aufhören müsse, Bewegung zu seyn” (ibid. p. 239).

[21] “An die Aufstellung von fünf Hauptformen der physischen Kraft reiht sich die Aufgabe, die Metamorphosen dieser Formen durch fünfundzwanzig Experimente zu beweisen” (Mayer, 1845, p. 34). 25 corresponds to the number of possible transformations between the various forms, which gives 20 in total, plus 5 arising from changes with maintenance of form.

new compared to the article of 42 is the generalization of the concept of force to the organic domain.[22] New is also the classical dictum:"*Ex nihilo nil fit. Nil fit ad nihilum*" (Nothing comes from nothing. Nothing becomes nothing) (ibid. p. 5). This statement is, however, equivalent to '*causa aequat effectum*'. As in 1842, the cause is said to be force and so is the effect. The force remains in quantity and only the form can vary.[23] Let us move on to phenomena.

Motion into motion

A white ball collides frontally and elastically with a red ball. The white ball loses its motion and the red ball moves on with the velocity of the white one.[24] The interpretation given is as follows: the motion of the white ball was the one that spent and originated that of the red one, it transformed into the latter.[25] The conservation of the quantity of force is justified in the following terms: the motion of the white ball is a force; the motion of the red ball is its effect; therefore, it is the same as the motion of the white ball. Thus, there was transformation, from motion into motion, and the quantity of force remained the same. In this way Mayer subsumes the conservation of the living forces for the elastic collision in his theory.[26]

The Leibnizian principle of living forces is subsumed as in the 1842 paper (Mayer, 1845, pp. 7–9). The relationship between motion and heat is treated anew in this book but with more detail than in the 42 paper. The calculation which had led to the value of the mechanical equivalent of heat, is now made explicit as follows.

Heat and motion

The experimental basis is the same as in the paper addressed above. To increase the temperature of a given quantity of gas without changing its volume, we need a certain amount of heat x, says Mayer (Fig. 2.2). In order to obtain the same temperature increase without changing the pressure, (allowing for a change in volume) a greater amount of heat is needed, be it $x + y$.[27] Since in one case there was a volume variation and in the other there was not, and the amounts of heat used for the same temperature

---

[22] " *Es giebt in Wahrheit nur eine einzige Kraft.* In ewigem Wechsel kreist dieselbe in der todten wie in der lebenden Natur. Dort und hier kein Vorgang ohne Formveränderung der Kraft!" (ibid. p. 6).

[23] "Die Wirkung ist gleich der Ursache. Die *Wirkung* der Kraft ist wiederum Kraft. Die *quantitative* Unveränderlichkeit des Gegebenen ist ein oberstes Naturgesetz, das sich auf gleiche Weise über Kraft und Materie erstreckt" (ibid. p. 5).

[24] "Stösst der weisse Ball den Rothen central an, so verliert der Weisse seine Bewegung und der Rothe geht mit dessen Geschwindigkeit fort" (ibid. p. 7).

[25] "Die Bewegung des Weissen ist es, welche aufgewendet die Bewegung des Rothen hervorgebracht, oder sich in die letztere *verwandelt* hat" (ibid. p. 7).

[26] "Die *Grösse* der Kraft aber, oder die sogenannte "lebendige Kraft der Bewegung" ist vor und nach dem Stosse constant geblieben" (ibid. p. 7).

[27] "Angenommen, ein Kubikzoll Luft von 0° und 27 Zoll Quecksilber Druck, sey durch die Wärmemenge x bei constantem Volumen um 274 °C. erwärmt worden [...] Ein andermal aber werde unser Cubikzoll Luft nicht unter constantem *Volumen*, sondern unter constantem *Drucke* der 27zölligen Quecksilbersäule von 0 auf 274° erwärmt. Diessmal ist eine grössere Wärmemenge erforderlich als zuvor; es sey dieselbe = x + y" (ibid. p. 12).

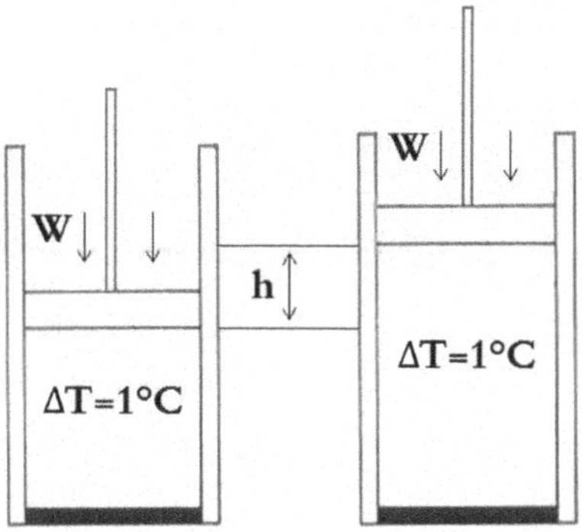

Fig. 2.2 Illustration of the experimental set up on which Mayer based his calculations. In both recipients, the temperature of the air increases by 1 °C. In the first case, the volume is constant, in the second, the pressure is constant. (The height *h* is not as large as represented.)

increase are in one case $x + y$ and in the other case only x, Mayer relates the difference between the heats used in the two cases, that is y, to the mechanical effect produced.[28] The determination of the value of y is based on experimental results of the time.

Let us move on to the figures. For calculation purposes, Mayer uses 1 cm^3 of atmospheric air at 0 °C and at the pressure of 76 cm of mercury. Heating the gas from 1 °C at constant pressure leads to a volume increase of 1/274 of the initial volume, which in this case means 1/274 cm.

As the gas lifts the air above the piston and this air balances a column of mercury 76 cm high (Fig. 2.3), the weight supported by the gas is equal to that of this column of mercury, i.e., 1033 g.

The product of the weight lifted by the height (falling force) is equal to

$$1033 \times \frac{1}{274} = 3.77\,\mathrm{g\,cm}$$

This mechanical effect is caused by the difference between the heat necessary to increase the temperature at constant pressure $(x + y)$ and constant volume (x). The value of this difference (0.000347–0.000244) is 0.000103° of heat. Now, if 0.000103° of heat corresponds to 3.77 g cm, then 1° corresponds to a given value

$$\frac{3.77}{0.000103} = 36,602 \frac{\mathrm{g\,cm}}{\textit{thermal unit}}$$

Mayer says, 1° corresponds to the mechanical effect of a gram at a height of 367 m.[29]

---

28 "Bei der Vergleichung dieser Vorgänge sehen wir in beiden die Luft von 0 auf 274° sich erwärmen und zugleich von einem Volumen auf zwei Volumina sich ausbreiten; im ersten Falle war die erforderliche Wärmemenge = x, im zweiten = x + y; im ersten Falle war der gelieferte mechanische Effekt = 0, im zweiten = 15 [Gewicht] [...] und 1" Höhe" (ibid. p. 12).

29 "Ein Kubikcentimeter atmosphärische Luft bei 0° und 0^{m},76 Barometer, wiegt 0.0013 Gramme; bei constantem Drucke um 1 °C. erwärmt, dehnt sich die Luft um 1/274 ihres Volumens aus und hebt somit eine Quecksilbersäule von einem Quadratcentimeter Grundfläche und 76 cm Höhe um

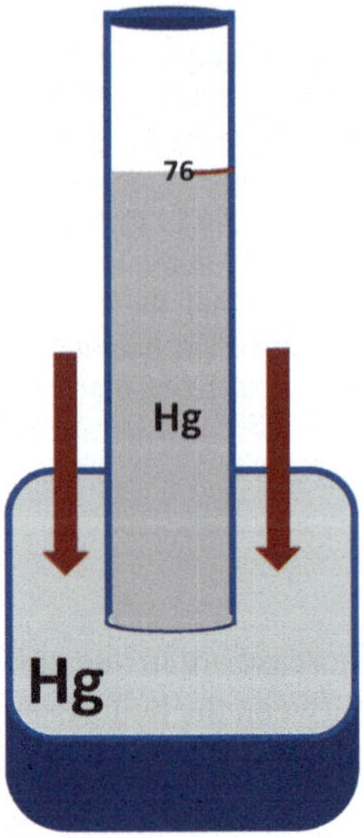

Fig. 2.3 The density of the mercury is 13.6 and the volume of the column is 76 cm^3. The product of volume and density is 1033.6 g

To give the result in actual units, we have to change centimeters to meters (366.02 g m), grams to kilograms (0.36602 kg m) and multiply by the acceleration of gravity ($g = 9.8$ m s^{-2}) to get weight. This gives 3.59 J. If we use the value given by Mayer, we obtain $0.367 \times 9.8 = 3.6$ J.

Electricity and motion

Let us move on to the connection between electricity and mechanical force. This connection is exemplified by an electrophorus. This is the apparatus pictured in Fig. 2.4, consisting of a base, a conducting plate and an insulating grip. The base could be made of glass or resin (Appendix E) and must be electrified by friction.

The upper part of the electrophorus, the plate, produces an electrical effect under certain conditions (if we bring a finger close, we get an electric shock). By lifting the plate, we can obtain a new effect (another electric shock). By letting the upper part go down to its initial position, we get another electric effect, and so on. Mayer puts forward the following reasoning. If the base is not electricized, the weight of the plate has a certain value *P* (Fig. 2.5a). If the base is electricized, it attracts the plate.

1/274 cm. Das Gewicht dieser Säule beträgt 1033 Gramme. Die specifische Wärme der atomsphärischen Luft ist bei constantem *Drucke*, die des Wassers = 1 gesetzt, nach *Delaroche* und *Bérard* = 0.267; die Wärmemenge, die unser Kubikcentimeter Luft aufnimmt, um bei constantem *Drucke* von 0 auf 1° zu kommen, ist also der Wärme gleich, durch welche 0.0013×0.267 oder 0.000347 Gramme Wasser um 1° erhöht werden. Nach *Dulong* […] verhält sich die Wärmemenge, welche die Luft bei constantem Volumen aufnimmt, zu der bei constantem Drucke, wie 1:1,421; hiernach gerechnet ist die Wärmemenge, die unseren Kubikcentimeter Luft bei constantem *Volumen* um 1°erhöht, $= \frac{0.000347}{1.421} = 0.000244$ Grad. Es ist folglich die Differenz $y = 0.0003470 - 0.000244 = 0.000103$ Grad Wärme, durch deren Aufwand das Gewicht P = 1033 Gramme auf $h = 1/274$ cm, gehoben wurde. Durch Reduktion dieser Zahlen findet man nun 1° Wärme = 1Grm. auf 367 ^{m} […] Höhe" (ibid. pp. 14–5).

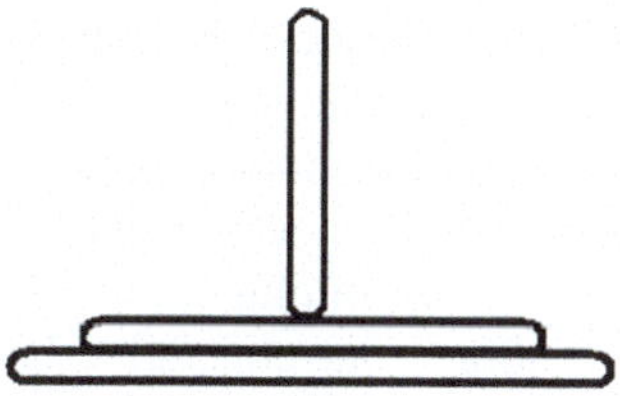

Fig. 2.4 Schema of an electrophorus

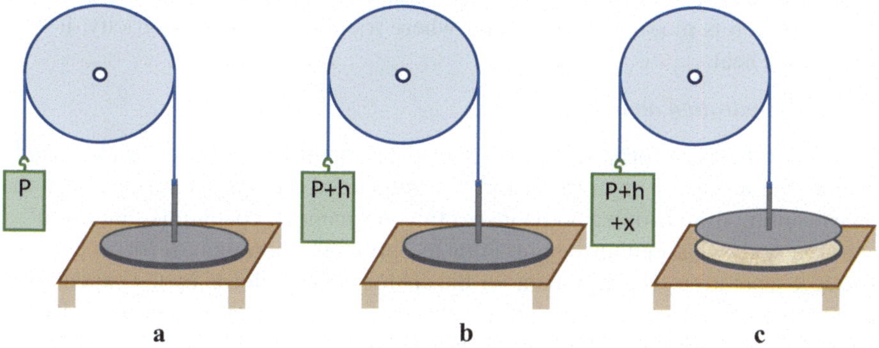

Fig. 2.5 **a** The base is not electricized; **b** the base is electricized (it attracts the plate); **c** the plate is lifted

To balance it in this case, we need a larger counterweight, be it $P + h$ (Fig. 2.5b). Now, we can get an electric effect, z. To lift the plate, we need a force greater than $P + h$, say $P + h + x$ (Fig. 2.5c).

When raised, another electrical effect, z', can be obtained. In each process (down—electric shock, up—electric shock) a force x is expended and effects z and z' are obtained. Since a force x is expended and an electrical effect $z + z'$ is gained, Mayer puts[30]

$$x = z + z' \tag{2.1}$$

In his terminology, the mechanical effect turned into electricity.[31] The argument is as follows:

---

[30] "[…] Auf der Unterscheibe liegend ist der Deckel im Stande einen elektrischen Effekt auszuüben; dieses ist geschehen, derselbe ist bestimmt worden und = z gefunden. Jetzt ist die Anziehung noch verstärkt und zur Hebung des Deckels bedarf es *eines noch grösseren* Gegengewichtes; das Produkt desselben in seine Höhe wird > Ph + p; es sey = Ph + p + x. Auf h erhalten wir den zweiten el. Effekt z' u.s.f. Bei jeder Senkung ist nun das gewonnene Produkt = Ph + p, bei jeder Erhebung aber das verlorene Produkt = Ph + p + x. *Während wir also jedesmal einen mechanischen Effekt = x aufwenden, gewinnen wir den el. Effekt z + z'*. So ist folglich: x = z + z'" (ibid. pp. 23–4).

[31] "*der mechanische Effekt ist in Elektrizität verwandelt worden*" (ibid. p. 24).

- the electricity in the lower part of the electrophorus remains constant, so it cannot have given rise to the electrical phenomena;
- on the other hand, a mechanical effect took place;
- therefore, either one admits that the mechanical effect had no consequence and the electrical effects came out of nothing—a 'double paradox'—or, admitting that 'nothing comes from nothing', one concludes that the mechanical effect was transformed into electricity.[32]

Another example of the transformation of mechanical effect into electricity is given by electricity obtained by friction.[33] Mayer remarks that in such a case, the heat of friction is missing.[34] Therefore, where friction produces electricity, it does not produce heat.

Chemical separation and junction

The fifth of the main forms of force is said to be chemical difference or chemically separate existence.[35] This form of force is introduced through an analogy with the force resulting from lifting a body: just as the 'mechanical separation of the earth'—that is, a body at a certain height from the ground—represents a force, so does chemical separation. The argument for the analogy is as follows: by the expenditure of the force of mechanical separation—thus by the fall—heat is produced; by the joining of some substances heat is also produced.[36]

One of the examples given is that of the joining of 1 g of hydrogen with 8 g of oxygen. The developed heat, known at the time (34,743°) is compared to the heat resulting from a falling body. It is equivalent to the heat resulting from a falling body 'weighing' two grams, coming from the limit of the action of attraction to the ground (34,700°).[37]

---

[32] "Aus Nichts wird Nichts. Die Elektrizität des Harzkuchens kann, da sie sich unvermindert erhalten hat, die fortlaufende Summe el. Effekte nicht hervorgebracht haben; der bei jedem Turnus verschwundene mechanische Effekt kann nicht zu Null geworden seyn. Was bleibt übrig, wenn man sich nicht in einem doppelten Paradoxon gefällt? nichts, als auszusprechen: *der mechanische Effekt ist in Eletrizität verwandelt worden*" (ibid. p. 24).

[33] "Die Erzeugung der *Reibungs-Elektrizität* erfolgt ebenfalls unter dem Aufwande von mechanischem Effekt" (ibid. p. 25).

[34] "Bekannt ist auch, dass bei der Bildung von Reibungs-Elektrizität die Reibungswärme fehlt" (ibid. p. 25).

[35] "Das chemisch-getrennt Vorhandenseyn, oder kürzer: *die chemische Differenz der Materie ist eine Kraft*" (ibid. p. 28).

[36] "Den *räumlichen* Abstand der Masse, *in specie* der Erde und eines Gewichtes, haben wir oben als eine Kraft kennen gelernt. Ein Gramme-Gewicht in unendlicher Entfernung—oder wie wir kürzer sagen wollen: *in mechanischer Trennung* von der Erde, stellt eine Kraft dar; durch den Aufwand dieser Kraft, d. h. durch die mechanische Verbindung beider Massen, wird eine andere Kraft erzeugt: die Bewegung eines Gramme-Gewichtes mit der Geschwindigkeit von 34450'; durch den Aufwand dieser Bewegung lässt sich ein Gramme Wasser um 17356° erwärmen. Die Erfahrung lehrt nun, dass derselbe Effekt, wie bei der *mechanischen* Verbindung, eine Wärmeentwicklung nemlich, erzielt wird durch die *chemische* Verbindung gewisser Materien" (ibid. pp. 26–8).

[37] "Die chemische Verbindung von 1 Gramme Wasserstoff (die Verbrennungswärme desselben nach *Dulong* = 34743° angenommen) mit 8 Gramme Sauerstoff ist *äquivalent* der mechanischen

Organic phenomena

Most of the 1845 book is devoted to living beings. Organic phenomena are treated analogously to inorganic ones, as we shall see. Plants perform a given activity, but only if they have sunlight. Thus, their activity does not arise from nothing.[38] In the case of animals, continues Mayer, the chemical force of food and the oxygen absorbed by breathing is at the origin of the movements they perform and the heat they develop.[39] This would justify why a person in activity needs more chemical force than a person at rest. Mayer gives numerical examples from studies carried out at the time: the amounts of feed for horses at rest and in motion were different[40]; the feeding of prisoners, who are at rest, soldiers in the barracks or workers under great physical exertion differed.[41] The available data therefore showed that with increased activity more chemical force is needed.

The similitude in the treatment of the organic and inorganic may be sharper, as the following. Just as the heat supplied to a gas at constant pressure, say x, serves to increase the temperature of y and have a mechanical effect of z, so the heat corresponding to the oxidation process taking place in muscles, say x', is equivalent to the heat released, y', and the mechanical effect produced, z'. In one case, therefore, it holds

$$x = y + z \tag{2.2}$$

Verbindung von 2 Gramme Gewicht mit der Erde; die Wärmeentwicklung bei beiden ist = 34700°" (ibid. p. 28).

The meaning of 'coming from the limit of action of attraction to the ground' is the following: "Der Begriff einer unendlichen Entfernung ist hier im physischen und nicht im mathematischen Sinne zu nehmen, und unter demselben "die physische Grenze der Anziehungssphäre" der Erde zu verstehen [...] Setzt man beispielsweise statt einer unendlichen Entfernung von der Erde eine von 10000 Erdhalbmessern, so genügt eine solche für die hier betrachteten Fälle vollkommen" (ibid. p. 27).

[38] "Die Erschaffung einer physischen Kraft, schon an und für sich selbst kaum denkbar, erscheint um so paradoxer, wenn man die Erfahrung berücksichtigt, dass die Pflanze einzig mit Hülfe des Sonnenlichtes ihre Leistung zu vollbringen im Stande ist" (ibid. p. 40).

[39] "Die chemische Kraft, welche in den eingeführten Nahrungsmitteln und in dem eingeathmeten Sauerstoffe enthalten ist, ist also die Quelle *zweier* Kraftäusserungen, der Bewegung und der Wärme, und die *Summe* der von einem Thiere producirten physischen Kräfte ist gleich der Grösse des gleichzeitig erfolgenden chemischen Processes" (ibid. pp. 45–6).

[40] "Ein starkes Pferd, das Tag für Tag der Ruhe pflegen darf, wird mit 15 [...] Heu und 5 [...] Hafer reichlich genährt; hat aber jezt das Thier, wie oben angenommen wurde, täglich 12′960000 [...] 1' hoch zu heben, so kann es bei dieser Nahrung offenbar nicht bestehen. Wir legen ihm, um es in gutem Stande zu erhalten, 11 [...] Hafer [...] zu" (ibid. p. 51).

[41] "Nach *Liebig* [...] erhalten die Gefangenen im Arresthause in Giessen, denen jede Bewegung mangelt, täglich 17 Loth (64 Lth. = 1 Kil.) Kohlenstoff. [...] Ein kasernirter Soldat geniesst täglich [...] 29 Loth Kohlenstoff. Gönnen wir aber unserem Arbeiter zur Vollbringung seiner schweren Leistung noch weitere 8 Loth, so wird er täglich 36 Loth einführen [...] Davon verwendet er zu mechanischem Effekte [...]" (ibid. p. 52).

and in the other[42]

$$x' = y' + z' \tag{2.3}$$

In another passage, the elasticity of a gas is analogized to the irritability of muscles: without heat, the gas has no elasticity; and without a chemical process, there is no irritability. The reason for this analogy is this: where nothing is, nothing can be transformed either.[43]

Conclusion

In dealing with phenomena, Mayer establishes a causal relationship between observables; and establishes an equivalence between the quantities that express the terms of this relationship. If he does not have numerical data, the equivalence is expressed formally (as in the case of the electrophorus). The justification for equivalence is given by the principle 'nothing comes from nothing and nothing becomes nothing'.[44] This principle being valid, one cannot admit '*a plus or a minus*' in any of the members of a cause-effect relationship, says Mayer.[45] The approach to biological processes is based on this same principle. It implies, he claims, that there is only transformation of force or matter, but never *creation.*[46]

---

42 "Wenn zu einer unter constantem Drucke sich befindenden Gasart eine bestimmte Menge von Wärme = x hinzutritt, so wird ein Theil dieser Wärme zur Temperaturerhöhung des Gases verwendet, und dieser Theil = y besteht als freie Wärme fort, ein anderer Theil wird "latent" und bringt den mechanischen Effekt = z hervor. Es ist nun x = y + z.

Setzen wir den in den Cappilaren eines Muskels vor sich gehenden Oxydationsprocess, oder die diesem entsprechende Wärme = x', die wirklich entwickelte freie Wärme = y', und den gelieferten mechanischen Effekt = z', so ist wieder x' = y' + z'" (ibid. pp. 98–9).

43 "Da die Elasticität der Gase und die Irritabilität der Muskeln Eigenschaften sind, die sich auf die Metamorphose gegebener Kräfte beziehen, so ist die Existenz dieser Eigenschaften nothwendig an die Existenz der respectiven Kräfte geknüpft. Wo nichts ist, da lässt sich auch nichts umwandeln. Ohne Wärme ist keine Elasticität, ohne chemische Differenz, oder ohne chemischen Process, keine Irritabilität denkbar" (ibid. p. 100).

44 "Aus dem Angeführten geht zur Genüge hervor, dass die höchst werthvollen Versuche der genannten Naturforscher, weit entfernt, eine Widerlegung des Grundsatzes: *ex nihilo nil fit* zu enthalten, vielmehr die angefochtene Wahrheit auf dem Erfahrungswege bestätigen" (ibid. p. 49). "Ein gewandter Schmied bringt ein kaltes Stück Eisen durch Hämmern in's Glühen; diese Wärme aber entsteht auf Kosten der Temperatur seines Armes. *Ex nihilo nil fit*" (ibid. pp. 95–6).

45 "Sammelt man die in einer gewissen Zeit von einem Thiere gelieferten mechanischen Kraftäusserungen, verwandelt dieselben [...] in Wärme, und addirt hiezu die in gleicher Zeit von dem Körper unmittelbar entwickelte Wärme, so wird man genau die Wärmemenge erhalten, welche dem stattgehabten chemischen Processe an und für sich entspricht. Auf der einen oder der anderen Seite ein *Plus* oder *Minus* anzunehmen, verbietet das Gesetz des logischen Grundes *Ex nihilo nil fit; nil fit ad nihilum*" (ibid. p. 46).

46 "Der Verfasser glaubt daher auf das Einverständniss seiner Leser rechnen zu dürfen, wenn er der folgenden Untersuchung als axiomatische Wahrheit den Satz unterlegt: *dass während des Lebensprocesses nur eine Umwandlung, so wie der Materie, so der Kraft, niemals aber eine Erschaffung der einen oder der anderen vor sich gehe*" (ibid. p. 40).

'Transformation' is the term generally used, although 'metamorphosis' also appears, with the same sense, sometimes even with the same referent.[47] These terms do not explain, however, the underlying natural process. He stresses, the transformation of chemical force into mechanical effect, which is manifested in the action of muscles, occurs in a *secret way to us*.[48] With 'transformation of heat into mechanical effect' a 'fact' is expressed, he says, but in no way did he explain the transformation. He clarifies this with an example.

A piece of ice turns into water. This is a 'fact'. It is a fact of observation: we had ice and now we have water. The observation fact, he adds, does not depend on questions about *how and why* this happens. Therefore, by saying that ice turned into water, one is not adding something about the melting of ice that is not observable. However, there is more. Mayer distances himself from such questions. He says that these are fruitless questions, proper to poets and philosophers of nature.[49] Thus, according to what he said about transformation and the dichotomy between positive knowledge and matters of fantasy, Mayer distances himself from an explanation of phenomena. Therefore, transformation does not explain the natural processes involved. The term is used to relate observables, which in phenomena are viewed as cause and effect.

2.1.2.1 Concerning the Principle of Conservation of Energy

The principle of conservation of energy tells us that energy can neither be created nor destroyed but only transformed. We may understand that an entity that cannot be destroyed or created must exist for ever. As energy is this entity then energy is conserved. Thus, the principle of conservation of energy justifies this conservation through its content.

Mayer formulated the principle, which he calls *axiom*, as follows:

> Bei allen physikalischen und chemischen Vorgängen bleibt die gegebene Kraft eine constante Grösse (ibid. p. 32)
>
> 'In all physical and chemical processes, the given force remains a constant magnitude.'

If we replace force by energy, only to make the proposition more accessible to comparison, Mayer's principle states:

---

47 When for example the 25 experiments for the five main forms of forces are referred to, the experiments are presented as metamorphoses (Metamorphosen) and in the specification of the experiments appears transformation (Umwandlung, Verwandlung) (ibid. pp. 34–5).

48 "Die Action des Muskels, die Umwandlung von chemischer Kraft in mechanischen Effekt, wird auf geheimnissvolle Weise durch einen Contact-Einfluss bedingt, der erfahrungsgemäss dem Nervensysteme zukommt" (ibid. pp. 106–7).

49 "Wenn hier eine Verwandlung der Wärme in mechanischen Effekt statuirt wird, so soll damit nur eine Thatsache ausgesprochen, die Verwandlung selbst aber keineswegs erklärt werden. Ein gegebenes Quantum Eis lässt sich in eine entsprechende Menge Wassers verwandeln; diese Thatsache steht fest da und unabhängig von unfruchtbaren Fragen über Wie und Warum und von gehaltlosen Speculationen über den letzten Grund der Aggregats-Zustände. Die ächte Wissenschaft begnügt sich mit positiver Erkenntniss und überlässt es willig dem Poëten und Naturphilosophen, die Auflösung ewiger Räthsel mit Hülfe der Phantasie zu versuchen" (ibid. p. 10).

'In all physical and chemical processes, *energy is a constant magnitude.*'

If energy is a constant magnitude, then it can be said that energy is conserved. The principle of the conservation of energy based on Mayer's axiom would then state: 'energy is conserved'. In this form, there is no entity with the properties of energy that justify that energy is conserved. Consequently, there is no ontological issue.

In sum, the principle of conservation of energy tells us why energy is conserved. As this is due to an entity called energy, we are led to ask the question of what this entity is. Mayer's principle tells us that energy is conserved but it does not justify this content by means of the being of energy. Without this being, we do not ask about it.

2.1.3 1848: Solar Heat and the Equivalent

In 1848, Mayer publishes a second book at his own expense, *Contributions to Celestial Dynamics. A Popular Presentation.* The subject of the book is the origin of solar heat. Mayer defends the thesis that solar heat comes from falling bodies on the surface of the Sun. The argument develops as follows. First, he considers the possible sources of heat. Second, he estimates the magnitude of the heat coming from the sun. On the basis of this value, he eliminates the possible sources of heat one by one, leaving the falling bodies at the end. In this explanation, the mechanical equivalent of heat plays a crucial role. It allows him to go from a mechanical phenomenon to a quantity of heat.

If there were solar heat emission without renewal, there would be a cooling of the star of 9000 °C in 5000 years, according to Mayer's calculations.[50] Since this does not occur, there will have to be renewal of the heat. Renewal could happen chemically or mechanically.[51]

Exclusively by chemical means—the sun being considered as a great heap of coal—Mayer continued, the star could not emit for more than 46 centuries.[52] Once the chemical path had been excluded, the mechanical path remained. Renewal could

---

[50] "Wird nemlich die Wärme-Capacität der Sonnenmasse, dem Volumen nach gerechnet, gleich der des Wassers [...] und denkt man sich zugleich den Wärmeverlust, den die Sonne durch Strahlung erleidet, auf ihre ganze Masse gleichförmig vertheilt, so ergibt sich für dieselbe eine jährliche Abkühlung = 1°, 8, wonach die Sonne in der geschichtlichen Zeit von 5000 Jahren eine Temperaturabnahme von 9000° erlitten haben müsste" (Mayer, 1848, p. 8).

[51] "Als algemeines Naturgesetz, von dem keine Ausnahme statt findet, gilt der Satz: dass zur Erzeugung von Wärme ein gewisser *Aufwand* erforderlich ist. Dieser Aufwand, so verschiedenartig er sonst seyn mag, lässt sich immer auf zwei Hauptkategorien zurückführen; es besteht derselbe nemlich entweder in einem chemischen Material oder in einer mechanischen Arbeit" (ibid. p. 3).

[52] "Liegt dieser Wiederersatz in einem chemischen Processe? Nehmen wir, um dieser Vermuthung so viel nur möglich einzuräumen, die ganze Sonnenmasse für einen Klumpen Steinkohlen, wovon jedes Kilogramm 6000 Wärmeeinheiten durch Verbrennung liefert, so wäre die Sonne nicht weiter als 46 Jahrhunderte lang im Stande durch ihren Brand den genannten Wärmeaufwand zu bestreiten" (ibid. p. 8).

then happen by friction, by the living force of the sun's rotation, or by collision, the falling of bodies on the sun. The hypothesis of renewal by friction is excluded because Saturn rotates faster than the sun and emits no such heat.[53] The second hypothesis is also ruled out. The living rotational force of the sun would only cover 183 years of heat expenditure.[54] Finally, the third hypothesis holds: the renewal of solar heat comes from falling masses attracted by the sun.[55]

The argument in favor of this thesis is based on observations available at the time and the mechanical equivalent of heat. First, Mayer argues that the number of celestial bodies—planets, moons, comets and asteroids—is inestimably large. In this regard he recalls a statement by Kepler, 'there are more comets in the sky than fish in the ocean'[56] and the large number of the bodies named asteroids by Arago.[57] Then, he uses a fact from astronomical observations: there are bodies, whose path diameter around the sun decreases. Mayer infers, that the smaller the bodies, the faster their trajectory decreases,[58] and then concludes that heavy substances fall on the solar surface.[59]

---

[53] "Es wurde die Vermuthung ausgesprochen, die Axendrehung der Sonne könnte das ursächliche Moment von ihrem Strahlen seyn [...] Einen raschen Umschwung für sich allein, ohne Reibung, ohne Widerstand, kann man sich nicht als die Ursache einer Licht- und Wärmeentwicklung denken, zumal da die Sonne sich keineswegs durch ihre Umdrehungsgeschwindigkeit vor den übrigen Körpern des Planetensystems auszeichnet [...] Der äussere Ring des Saturns übertrifft den Sonnenäquator in seiner Rotations-Geschwindigkeit um mehr als das zehnfache. Nichts destoweniger wird aber weder an der Erde, noch am Jupiter, noch am Saturnusringe eine Licht- und Wärmeerzeugung wahrgenommen" (ibid. pp. 8–9).

[54] "[...] so geht aus dem bisherigen hervor, dass der ganze Rotations-Effect der Sonne, wenn durch ihn der Wärmeverbrauch gedeckt werden sollte, in hundert drei und achtzig Jahren verzehrt seyn müsste" (ibid. p. 10).

[55] "[...] so haben diese wandernden Himmelskörper in der Peripherie des Sonnensystemes ihre Wiege, im Centrum ihr Grab [...] Alle diese Massen stürzen mit einem heftigen Stosse in ihr gemeinsames Grab. Da nun keine Ursache ohne Wirkung besteht, so muss auch jede dieser kosmischen Massen, ebenso wie ein zur Erde fallendes Gewicht, durch ihren Stoss eine, ihrer lebendigen Kraft proportionale Wirkung, eine gewisse Menge von Wärme, hervorbringen" (ibid. p. 12).

[56] "Im Raume des Sonnensystemes bewegen sich ausser den jetzt bekannten 14 Planeten und deren 18 Trabanten noch eine sehr grosse Zahl anderer Himmelsmassen, von welchen die Kometen zunächst Erwähnung verdienen.

Kepler's berühmter Ausspruch "es gibt mehr Kometen im Himmelsraume, als Fische im Ocean" gründet sich auf die Thatsache [...]" (ibid. p. 13).

[57] "Neben den Planeten, Monden und Kometen gibt es aber in unserem Sonnensysteme noch eine weitere Kategorie von Himmelskörpern. Es sind dieses geballte Massen [...] denen *Arago* den passenden Namen "Asteroiden" gegeben hat. Wie die Planeten und Kometen, so folgen auch sie den Gesetzen der Schwere und umkreisen in elliptischen Bahnen die Sonne" (ibid. p. 14).

[58] "Es ist der Natur der Sache angemessen, dass die Planeten, ihrer ausnehmenden Grösse und Dichtigkeit wegen, eine nur sehr langsame und bis jetzt unmerkliche Verminderung ihrer Bahndurchmesser erfahren. Dagegen müssen sich die kleinen kosmischen Massen unter sonst gleichen Umständen in dem Masse der Sonne rascher nähern, je kleiner ihr körperlicher Durchmesser ist" (ibid. p. 15).

[59] "Aus der grossen Anzahl kometarischer Massen und der Asteroiden, sowie aus dem Vorhandenseyn der Zodiakallicht-Materie einerseits und der Existenz eines widerstandleistenden Aethers

The effect of the impacts on the solar surface would depend on the mass of the bodies, but still on their velocity.[60] Mayer calculates the minimum and maximum velocity with which a body hits the solar surface. Using the mechanical equivalent of heat, he concludes then that an asteroid gives the sun between 4600 and 9200 more heat units per fall than an equal mass of coal.[61]

Another argument for Mayer's thesis is derived from the diathermic energy of radiation. Experiments had shown that the higher the temperature of the source, the greater the radiation that penetrates space.[62] From this he infers that the solar surface must be at a much higher temperature than is reached in combustion.[63] A heat of the order of magnitude of that calculated for the solar source, however, could be obtained from falling asteroids.[64] Mayer then calculates the amount of matter that should fall on the star per minute to be between 94,000 and 188,000 billion kilograms.[65] This might seem like a large figure, but given the solar surface area, it would represent an increase of 15–30 g per square meter.[66] Mayer then recalls that a light rain for one hour amounts to 17 g per square meter.[67]

---

andererseits ergibt sich mit Notwendigkeit, dass fort und fort wägbare Substanzen auf der Sonnenoberfläche anlangen müssen" (ibid. p. 16).

60 "Der Effect aber, den diese Massen dort ausüben, hängt offenbar von ihrer Endgeschwindigkeit ab" (ibid. p. 16).

61 "Nach der am Schlusse des 2ten Capitels gegebenen Formel ist der Wärme-erregende Effect beim Stosse = 0°,000139 × c^2, wo c die Geschwindigkeit des stossenden Körpers nach Metern gerechnet ausdrückt. Da nun die Geschwindigkeit eines Asteroids beim Zusammenstossen mit der Sonne 445 750 bis 630 400 m beträgt, so ist der Effect = 27 ½—bis 55 Millionen Grad Wärme. Eine Asteroid-Masse gibt also bei ihrem Sturze auf die Sonne 4600 bis 9200 mal so viel Wärme, als eine gleich grosse Menge Steinkohlen (à 6000 Calorien) durch Verbrennen liefert!" (ibid. p. 20).

62 "Dagegen wächst die diathermane Energie der Strahlen fortwährend, wie die Temperatur der Quelle höher wird" (ibid. p. 23).

63 "Dieses Erfahrungsgesetz, dass die Raum-durchdringende Energie der Wärmestrahlen im allgemeinen zunimmt, wenn die Temperatur ihrer Quelle eine höhere wird, lehrt, dass auf der Sonnenoberfläche eine weit grössere Hitze herrschen muss, als der heftigste Verbrennungsprocess hervorzubringen vermag" (ibid. p. 24).

64 "Vergleicht man diese künstlichen Feuer mit der von einem auf die Sonne stürzenden Asteroide hervorgebrachten Hitze, so findet man die letztere (auch abgesehen von der im Vergleiche zum Wasser wahrscheinlich ziemlich geringen Wärme-Capacität der Asteroid-Massen) 7- bis 14 000 mal grösser, als selbst die Hitze eines Knallgasgebläses, woraus sich denn die ausserordentliche diathermane Energie der Sonnenstrahlen, die unermesslich grosse Ausstrahlung der Sonnenoberfläche und die extreme Hitze im Focus der Brennspiegel leicht ergibt" (ibid. p. 25).

65 "Da nun [...] 1 K° Asteroid-Masse 27 ½—bis 55 Millionen Wärmeeinheiten gibt, so folgt daraus, dass die Quantität der auf die Sonne niederstürzenden kosmischen Materien in jeder Minute zwischen 94 000- und 188 000 Billionen Kilogramme betragen muss" (ibid. p. 29).

66 "Da nun das Quantum der in 1 min auf die Sonne stürzenden Asteroid-Massen 94 000-bis 188 000 Billionen Kilogramme beträgt, so folgt hieraus für 1 QuadratMeter Sonnenoberfläche eine Massenzunahme von durhschnittlich 15 bis 30 Grm. per Minute" (ibid. pp. 31–2).

67 "Um diesen Vorgang [...] erinnere man sich, dass ein schwacher Regen in einer Stunde eine 1 mm hohe Wasserschichte gibt (während starke Gewitterregen leicht das 10 bis 15 fache Quantum liefert) welches für 1 Quadrat-Meter Oberfläche in einer Minute 17 Grm. ausmacht" (ibid. p. 32).

Conclusion

Mayer defends the thesis that the heat of the sun originates from falling bodies on the solar surface. The mechanical equivalent of heat provides him with the tool of calculation, which allows him to go from a mechanical to a thermal value. In the remaining, he shows that it is plausible to admit that there is an innumerable set of bodies that fall on the sun; and that his explanation is more adequate than the alternatives.

2.1.4 1851: On the Mechanical Equivalent of Heat

In 1851, Mayer publishes a third book at his own expense, *Remarks on the Mechanical Equivalent of Heat*. This is a historical–philosophical defense of the mechanical equivalent of heat. He takes some scientific results from the past and shows that the essential thing in research was quantitative determination. Finally, he comes to the topic 'heat': it was not explanations of the nature of heat, but a number—the mechanical equivalent of heat—that improved knowledge of calorific phenomena.

The golden rule of research consists, according to Mayer, in trying to know the phenomena before trying to explain them by ultimate causes.[68] In order to know the phenomena, one must investigate until a numerical relationship can be determined. Herein lies the foundation of research, according to the author.[69]

Free fall is the first topic of the argument: the task of science was to establish a numerical relationship between height, time of fall, and final velocity. Experimental research led to the result.[70] The ascent of liquids by suction is another topic. It was not the explanations about the qualities of vacuum that led to the knowledge of the phenomenon, but the measurement.[71] The knowledge of combustion processes has

[68] "Die wichtigste, um nicht zu sagen einzige Regel für die ächte Naturforschung ist die: eingedenk zu bleiben, dass es unsere Aufgabe ist, die Erscheinungen *kennen* zu lernen, bevor wir nach Erklärungen suchen oder nach höheren Ursachen fragen mögen. Ist einmal eine Thatsache nach allen ihren Seiten hin bekannt, so ist sie eben damit erklärt und die Aufgabe der Wissenschaft ist beendigt" (Mayer, 1851, pp. 5–6).

[69] "Die Regel, nach welcher verfahren werden musste, um die Fundamente der Naturkunde in der denkbar kürzesten Zeit zu legen, lässt sich in wenige Worten fassen. Es müssen nämlich die nächstliegenden und häufigsten Naturerscheinungen mittelst der Sinnwerkzeuge einer sorgfältigen Untersuchung unterworfen werden, die so lange fortzuführen ist, bis aus ihr Grössenbestimmungen, die sich durch *Zahlen* ausdrücken lassen, hervorgegangen sind.

Diese Zahlen sind die gesuchten Fundamente einer exacten Naturforschung" (ibid. p. 7).

[70] "Unter allen Natur-Processen ist der freie Fall eines Gewichtes der häufigste [...] die Aufgabe besteht nun darin, die zwischen der Fallhöhe, der Fallzeit und der Endgeschwindigkeit stattfindenden Grössenbeziehungen aufzufinden und in *bestimmten Zahlen* auszudrücken. Bei der Ausführung dieser Experimental-Untersuchung [...]" (ibid. p. 7).

[71] "Eine zweite [...] alltägliche Erscheinung ist das Aufsteigen von Flüssigkeiten in Röhren beim Saugen. Auch hier gilt es wieder, sich nicht durch das *velle rerum cognoscere causas* zu nutzlosen und also schädlichen Speculationen über die Qualitäten des Vacuums u. d. gl. in die Irre führen zu lassen [...] Sollte etwa die Saugwirkung eine messbare Grenze haben? - Sobald wir einmal

also been numerically determined, this time by weighing the products before and after a process.[72] Finally the question, what about the heat developed by friction, inelastic collision or by pressure of gases?[73] Here too, the task was not accomplished with heat-substance, heat-aether, heat-atoms.[74] Rather, it was measurements of the work expended and the heat developed that made it possible to establish a relationship between the two.[75] The number, which establishes the relationship between the quantities of one and the other, is the mechanical equivalent of heat.[76]

Conclusion

Mayer's last book, on the mechanical equivalent of heat, highlights the role of the equivalent for scientific knowledge. First, when he shows that in several domains of science, it was the numerical relationship that brought progress. Second, when he refers to our conditions of knowing nature. He pointed out that our knowledge is limited: we do not know the real causes of phenomena; we do not know what force is; we know nothing about the essence of heat. The transformation of one force into another, which could reflect intricacies of phenomena, is explicitly understood as an expression of a mere numerical relationship.

---

anfangen, in dieser Richtung zu experimentiren, so kann es uns nicht mehr entgehen [...] Diese Zahl ist ein zweiter Hauptfeiler im Gebäude des menschlichen Wissens" (ibid. p. 8).

72 "Von jeher mussten die Verbrennungserscheinungen die Aufmerksamkeit der Menschen in besonderem Grade in Anspruch nehmen. Um sie zu *erklären*, stellten die Alten ihrer naturphilosophischen Methode gemäss ein besonderes, nach oben strebendes Feuerelement auf [...] Auch hier sind es Grössenbestimmungen, Zahlen allein sind es, die uns den Ariadne-Faden in die Hand geben. Wollen wir erfahren, was bei den Feuererscheinungen vorgeht, so müssen wir die Stoffe vor und nach ihrer Verbrennung *wägen* [...]" (ibid. pp. 11–2).

73 "Wir wissen aber längst, dass in einer Unzahl von Fällen Wärme auftritt, wo kein chemischer Process statt findet; so namentlich bei jeder Reibung, beim unelastischen Stosse und beim Zusammendrücken luftförmiger Körper" (ibid. p. 12).

74 "Die Geschichte lehrt, dass auch hier die scharfsinnigsten Hypothesen von dem Bestande und der Natur eines besonderen Wärme "stoffes", von einem bald ruhenden, bald schwingenden "Wärme-Aether", von "Wärme-Atomen", die in den zwischen den Massen-Atomen befindlichen Räumen ihre Rolle spielen sollten, u. s. w., die Aufgabe nicht zu lösen vermocht haben" (ibid. pp. 12–3).

75 "Es müssen wieder Grössenbestimmungen vorgenommen, es muss gemessen und gezählt werden. Wenn wir in dieser Richtung vorgehen und die auf mechanischem Wege entwickelte Wärmemenge, sowie die dazu verbrauchte Arbeitskraft messen, und diese Grössen mit einander vergleichen, so finden wir sofort, dass dieselben in der denkbar einfachsten Beziehung, d. h. in einem unveränderlichen, geraden Verhältnisse zu einander stehen [...] Diese Thatsachen in kurze, klare Worte gefasst, sagen wir: *Wärme und Bewegung verwandeln sich in einander*" (ibid. p. 13).

76 "das Gesetz der unveränderlichen Grössenbeziehung zwischen der Bewegung und der Wärme muss auch *numerisch* ausgedrückt werden. Indem wir die Erfahrung hierüber befragen, finden wir [...] *Diese Zahl ist das mechanische Aequivalent der Wärme*" (ibid. pp. 13–4).

2.1.5 Conclusion

Mayer determined the mechanical equivalent of heat in 1842 and 1845, used it in the explanation of the solar heat and wrote a book on it. This value is obtained from an equation of the form

$$\alpha \; mechanical \; units = \beta \; thermal \; units.$$

This equation, in turn, is obtained from a phenomenon for which a cause-effect relationship holds. This relationship, in general, is established by an observer of a phenomenon. With regard to electricity, Mayer arrives at the expression $x = z + z^0$ (where x stands for the mechanical part and $z + z^0$ for the electric). If he had the values of the mechanical and electrical parts, he could have written an equation in the form of

$$\alpha \; mechanical \; units = \gamma \; electric \; units.$$

On the basis of this, he could have calculated the mechanical equivalent of electricity, just as he did for the mechanical equivalent of heat. This comparison is legitimated since both heat and electricity are forms of force.

Mayer created a theory. In this theory, force is the central concept. This concept is related with the cause-effect relationship: in a phenomenon, cause and effect are called forces. The properties of force—indestructibility and transformability—stem also from this relationship (1842). He was aware that his theory needed an experimental basis. This basis was provided by the mechanical equivalent of heat, as he specially highlighted in the 1851 book.

2.2 James Joule

James Joule was not a physicist and had no academic training. His hobby was to perform experiments. In 1838, he published a paper in the journal *Annals of Electricity,* in which he presents a machine that produces electricity through motion (Joule, 1838).[77] In 1840, he establishes a quantitative relationship between electricity and heat.[78] The first Joule's paper that will interest us deals with experiments which involve motion and electricity, as in the 1838 apparatus, and electricity and heat, as the 1840 experiment. His focus is now on the relationship between *motion* and *heat.*

This relationship between motion and heat was important at that time due to the following. According to most physicists of the time, experiments had shown that

[77] Between 1838 and 1840, he worked on electromagnets and motors (Joule, 1884).

[78] About Joule's research on heat produced by electric currents in metals see Martins (2022) and Martins and Silva (2021).

the quantity of heat did not vary. (I.e., if one body gets hotter another gets colder.) If the quantity does not vary, then heat is something that subsists. For this reason, heat was taken as a substance. If heat is taken as a substance because its quantity does not vary, it follows that if the quantity of heat varies, heat cannot be taken as a substance. Therefore, if Joule showed that the amount of heat varies, heat would no longer be a substance. The prevailing thesis about heat would collapse. What would heat be if it were not a substance? The alternative at that time was to take heat as motion. In fact, neither the proponents of the thesis 'heat is motion' had observed the motion that heat was said to consist of nor the proponents of the thesis of heat as a substance had observed such a substance. Heat as a substance and heat as motion were interpretations of the experiments. Hence, the fundamental question was only whether the amount of heat is conserved or not, which is also the one that could have an experimental answer.

Joule tries to show through a series of experiments that the quantity of heat varies in phenomena so that heat cannot be a substance, but it is rather a kind of motion. The first experiments performed with this purpose were in 1843. In 1845, he conducted a series of experiments with gases, whose purpose was to show that there is a correlation between heat and motion. In this same year, Joule conducted water friction experiments. These experiments were retaken in 1847 and 1849 and became Joule's most famous experiment: the paddle wheel experiment. Again, Joule correlates the mechanical action with the developed heat.

2.2.1 1843: Motion, Heat and the Equivalent

The 1843 experiments were performed with a machine that had the peculiarity of producing electric current by means of the movement of a magnet. Since current is produced by moving the magnet, the machine was called a magneto-electric machine. This current was called magneto-electricity (Appendix H).[79] This current, in turn, produces heat. Therefore, the machine provides the sequence: motion, electricity, heat. What matters are the ends of the sequence. Electricity was just a medium.

The article begins by asking the question, whether the heat from the magneto-electricity arises by transfer or by generation.[80] This question is linked to another one: if heat is a substance or a state of vibration. For if heat is not a "substance" but a "state of vibration," says Joule, there is no reason why it should not be induced by a purely mechanical action, as he understood the magneto-electric machine to

[79] At that time, it was not known whether the electricity from this machine was like electricity from other sources, so they gave names to the electricity (Appendix H).

[80] "it must be admitted that hitherto no experiments have been made decisive of this very interesting question; for all of them refer to a particular part of the circuit only, leaving it a matter of doubt whether the heat observed was *generated*, or merely *transferred from the coils* in which the magneto-electricity was induced, the coils themselves becoming cold" (Joule, 1843, p. 123).

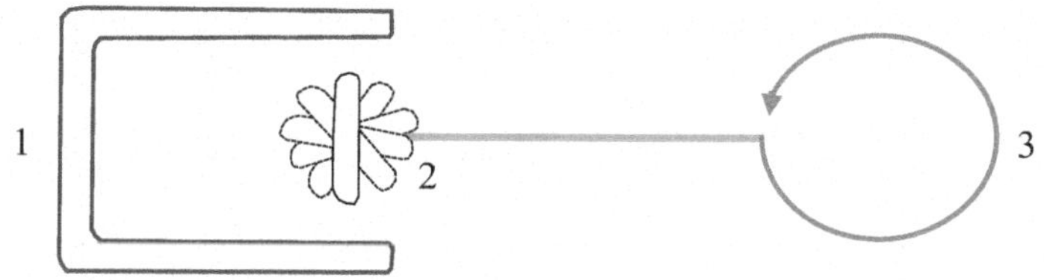

Fig. 2.6 Schema of the theoretically significant elements

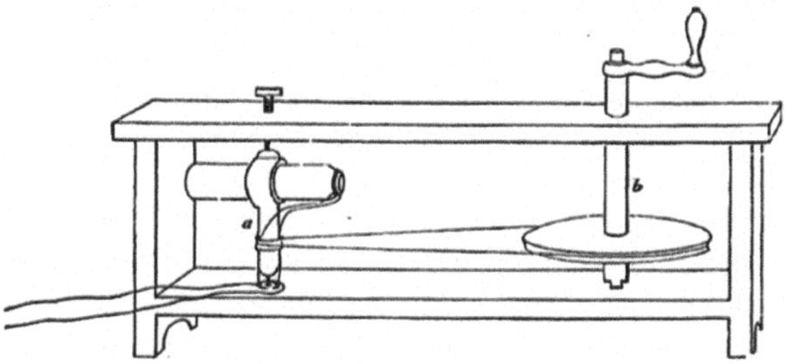

Fig. 2.7 Moving the crank, the axis (b) rotates and makes axis a to rotate. Perpendicular to axis a is an electromagnet immersed in a tube with water. This electromagnet moves close to an electromagnet, which is not shown (Joule, 1843, p. 125)

be.[81] From the various series of experiments in this study, Joule concluded that the magneto-electric machine can generate and destroy heat. Therefore, the amount of heat varies. If the quantity of heat varies, the question about the nature of heat is solved: heat cannot be a substance. Not being a substance, heat is motion.

Joule did not stop there. He intended to determine how much mechanical action is required to obtain one unit of heat. He then takes up some of the experiments performed in order to determine the *mechanical value of heat*. Let us move on to these experiments.

The magneto-electric machine consists of three relevant elements (Fig. 2.6):

1. a magnet or an electro-magnet[82];
2. a small electro-magnet in comparison with the previous one;
3. a crank that brings the smaller electromagnet into rotation. Due to this rotation, electrical current is developed.

Elements 2 and 3 are represented in Fig. 2.7.

---

81 "It is pretty generally, I believe, taken for granted that the electric forces which are put into play by the magneto-electrical machine possess, throughout the whole circuit, the same calorific properties as currents arising from other sources. And indeed when we consider heat not as a *substance*, but as a *state of vibration*, there appears to be no reason why it should not be induced by an action of a simply mechanical character, such, for instance, as is presented in the revolution of a coil of wire before the poles of a permanent magnet" (ibid. p. 123).

82 An electro-magnet is a device that works as a magnet but is driven by a battery.

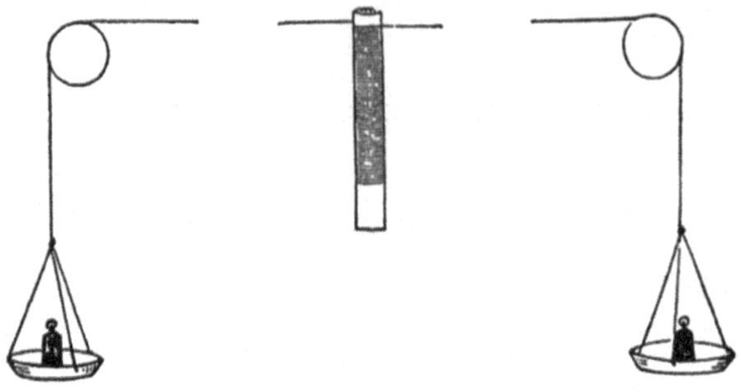

Fig. 2.8 The axis is wrapped by a double thread, which unfolds in diametrically opposite directions, each passing through a pulley and bearing weights at the ends (ibid. p. 150)

From the experiments carried out, Joule concluded that motion creates heat. As heat could be increased through the magneto-electric machine and this functions by means of motion, Joule tried to determine a proportion between the heat evolved and the mechanical power at stake. To estimate this mechanical power, he replaced the crank with the system represented in Fig. 2.8.[83]

The weight and height of the bodies make up the value of the mechanical power. This value is equated with the value of the heat evolved. Thus, he arrives at an equation of the form

$$\alpha \textit{ mechanical units} = \beta \textit{ thermal units}.$$

As he then knows that α mechanical units correspond to β thermal units, he is able to calculate how many mechanical units correspond to one thermal unit. This is the mechanical value of heat. Let us have a look at the figures.

The machine moves with a velocity of 600 rotations per minute. For this motion, a weight of 5 lb. 3 oz (2.35301 kg) in each scale is necessary. These falling bodies have two effects: the 600 rotations per minute and the current produced. Joule is interested in the heat that comes from electricity. To get it, he takes out of the total mechanical action what is spent to maintain the 600 rotations per minute. To obtain this latter value, he set the machine in motion with 600 rotations per minute without producing electricity. Under these conditions, it needs only 2 lb. 13 oz (1.27573 kg).[84] The difference between these two weights is then taken as the weight which contributes to the heat produced,

[83] "The axle *b* [...] was wound with a double strand of fine twine, and the strings [...] were carried over very easily-working pulleys, placed on opposite sides of the axle [...] By means of weights placed in the scales attached to the ends of the strings, I could easily ascertain the force necessary to move the apparatus at any given velocity" (ibid. p. 150).

[84] "when the battery was thrown out of communication with the electro-magnet, and the motion was opposed solely by friction and the resistance of the air, only 2 lb. 13 oz. were required for the same purpose" (ibid. pp. 150–1).

$$2 \times (5\,\text{lb.}\,3\,\text{oz}-2\,\text{lb.}\,13\,\text{oz}) = 4\,\text{lb.}\,12\,\text{oz}\,(2.15456\,\text{kg})$$

Each scale covers the distance of 517 feet (157.582 m). The mechanical power used is then equal to

$$4.75\,\text{lb.} \times 517\,\text{ft} = 2455.75\,\text{lb. ft}\,(339.519\,\text{kg m})$$

The final value of heat evolved is determined. It is equivalent to the heat necessary for heating a pound of water by 2.74°F (690.929 cal).[85] Thus, if 2455.75 lb.ft corresponds to 2.74°, 1° corresponds to *x*,

$$4.75\text{lb}.517\text{ft} \rightarrow 2.74^\circ$$
$$x \leftarrow 1^\circ$$

where[86]

$$x = \frac{4.75 \times 517}{2.74} = 896.259\,\text{lb.ft/BTU}.$$

Joule said that the mechanical value of heat is equal to the weight of 896 pounds which fall from the height of one foot.[87] Two other experiments gave the results of 1001 and 1040 lb.ft.[88]

The next series of experiments provides him with the reason for another thesis: heat can be destroyed through the magneto-electric machine. The experimental configuration that supports this statement differs from the previous because it includes a battery. (As the battery had been discovered by Volta, its current was called voltaic electricity.) This current generates heat. Joule determines the heat generated by this electricity alone and the heat generated when both kinds of electricity are involved. Comparing the results, he argues that the quantity of heat due to voltaic electricity can be decreased or increased when both kinds of electricity are involved. If it decreases, he claims that heat was destroyed; if it increases, heat was created. Let us see how he arrives at these conclusions.

Joule connects element 2 (Fig. 2.6) with a battery and measures the heat developed by this current, when no magneto-electricity is involved. In this case, the machine is at rest. To obtain this situation, one simply removes the large electromagnet (element

---

85 "2°.46 × 1.114 = 2°.74; and this has been obtained by the power which can raise 4 lb. 12 oz. to the perpendicular height of 517 feet" (ibid. p. 151).

86 About the conversion of units: 1 lb. = 0.453592 kg, 1 *ft* = 0.3048 m, 1 BTU = 252.164 cal. Therefore, $\frac{1\,\text{lb.}\cdot 1\,\text{ft}}{1\text{BTU}} = 0.000548\,\text{kg m/cal}$. Thus, 0.000548 × 9.8 = 0.00537 J/cal. Thus, multiplying the value of the mechanical equivalent of heat determined by Joule by 0.00537, we obtain this value in modern units.

87 "1° of heat per lb. of water is therefore equivalent to a mechanical force capable of raising a weight of 896 lb. to the perpendicular height of one foot" (ibid. p. 151).

88 "Two other experiments, conducted precisely in the same manner, gave a degree of heat to mechanical forces represented respectively by 1001 lb. and 1040 lb." (ibid. p. 151).

1, Fig. 2.6). When, in a second step, the electromagnet is added again, the machine starts to work by itself, as if it were an engine. When this motion is finished, the heat is determined. The heat developed is less than when the machine was at rest. From this Joule concluded that the magneto-electrical machine can destroy heat. Let us move on to the figures.

When the machine works as an engine,[89] there are two effects to consider: the rotational motion and the electricity produced. To separate these effects, Joule puts the machine in motion with the same number of revolutions per minute, without producing electricity. Thus, he determines the value of the mechanical effect. As in the previous cases, the product of the weights used by the height covered is taken as the mechanical power of the machine. This value was 2.25 lb. × 275 ft = 618.75 lb.ft.

The quantity of heat developed, when the machine works as an engine is of 1.191° calculated along the lines of the previous ones and also referred to a pound of water (ibid. p. 153). The quantity of heat developed in the circuit when the electromagnet is at rest is 1.794 (ibid. p. 153). As this value is higher than the previous one, Joule concludes that the difference (1.794–1.191 = 0.603) has been converted into mechanical power (ibid. p. 153). Now, since this heat difference corresponds to a movement that could be created by a mechanical power of 618.75 lb.ft, to the thermal unit corresponds

$$\frac{2.25 * 275}{0.603} = 1026.12\,\text{lb.ft/BTU}$$

Joule concludes that the drop of 1026 lb. from the height of a foot is the mechanical power required to raise the temperature of a pound of water 1 °F.[90] Another experiment carried out in the same way provides the value of 587 lb.ft (ibid. p. 153).

In these experiments, in which the machine works as an engine, the electromagnet rotates in one and the same direction. Joule will now force it to rotate in the opposite direction. To do this, he needs an external action. He uses again the falling bodies and follows the same steps as previously. In order to set the machine into motion now, 6 lb. 4 oz in each scale are required (ibid. p. 152)

$$W_{\text{total}} = 2 \times 6\,\text{lb.}\,4\,\text{oz} = 12.5\,\text{lb.}$$

As in the previous experiment, he takes from this total weight what is exclusively mechanical, to find out what can be attributed to heat. To do this he puts the machine to work as a mere mechanical machine. The weights necessary for that are 2 lb. 8 oz in each scale (ibid. p. 152). Therefore,

$$W_{\text{mech}} = 2 \times 2\,\text{lb.}\,8\,\text{oz} = 5\,\text{lb.}$$

89 "An experiment was now made, using the same apparatus as an electro-magnetic engine" (ibid. p. 153).

90 "Hence 0°.603 has been converted into a mechanical power equal to raise 2 lb. 4 oz. to the height of 275 feet […] one degree per lb. of water may be converted into the mechanical power which can raise 1026 lb. to the height of one foot" (ibid. p. 153).

The difference between weights W_{total} and W_{mech} (equal to 7.5 lb) indicates the weight related to heat production. As the scales cover the distance of 551 ft,[91] the mechanical power used in producing that quantity of heat is equal to

$$7.5\,\text{lb.} \times 551\,\text{ft} = 4132.5\,\text{lb.ft}$$

Let us consider the corresponding heat. Joule determines the heat developed and arrives at 9.92° (ibid. p. 152). When element 2 (Fig. 2.6) is at rest, the heat developed is equivalent to 5.38° Fahrenheit per pound of water. The difference between the heat in motion and at rest is therefore 9.92–5.38 = 4.54. Now, if at 4.54° corresponds to a mechanical power of 4132.5 lb.ft (ibid. p. 152), it follows that to one thermal unit corresponds

$$\frac{4132.5}{4.54} = 910.242\,\text{lb.ft/BTU}$$

Joule concludes that the heat required to raise a pound of water by 1 °F is equivalent to 910 lb. raised to the height of a foot.[92] A last series of experiments is performed with an iron cylinder, previously coated with copper as element 2 (Fig. 2.6), the revolving piece (ibid. p. 154). The heat generated and the mechanical power are determined as in the previous cases. From five experiments thus performed Joule concludes: 1° of heat per pound of water can be generated by the expenditure of mechanical power capable of raising 742 lb. to the height of one foot (ibid. p. 156). The figures are

$$\frac{4.2 * 517}{2.926} = 742.105\,\text{lb.ft/BTU}$$

Two other experiments conducted along the same lines, only with a less powerful battery, lead to 860 lb.ft (ibid. p. 156).

Gathering the obtained values, 13 in total (Fig. 2.9), Joule calculates the arithmetic mean

$$\frac{587 + 5 * 742 + 2 * 860 + 896 + 910 + 1001 + 1026 + 1040}{13} = 837.69\,(4.50\,\text{J})$$

and arrives at the value of 838 lb.ft as the mechanical power equivalent to the amount of heat capable of raising the temperature of a pound of water by 1°F (ibid. p. 156).

[91] "Hence 4°.54 were evolved in the experiment over and above the heat due to the chemical changes taking place in the battery, by the agency of a mechanical power capable of raising 7 lb. 8 oz. to the height of 551 feet" (ibid. p. 152).

[92] "one degree is equivalent to 910 lb. raised to the height of one foot" (ibid. pp. 152–3).

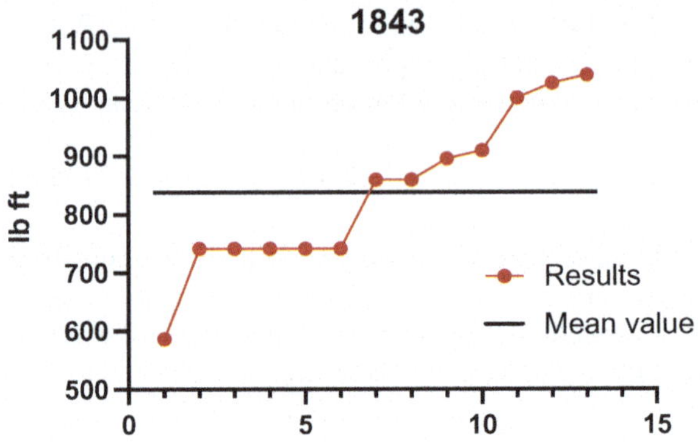

Fig. 2.9 The mechanical values of heat obtained by Joule presented in an increasing order.

Conclusion

Joule used falling bodies to determine the mechanical action. This action sets the machine in motion and produces heat. Joule wants to know the value of the action that produces heat. To do this, he takes the total of the action and subtracts that which serves to set the machine in motion as a mere mechanical object. From this comes

$$(weight \times height)_{total} - (weight \times height)_{mec.} = heat_{mag-elec.} \tag{2.4}$$

From such equations he calculates the mechanical value of the heat.

When the heat of magneto-electricity and voltaic electricity are at stake, he subtracts their respective values

$$heat_{mag-elec.} - heat_{voltaic\ elec.}$$

He did the same with regard to the mechanical action

$$(weight \times height)_{total} - (weight \times height)_{mec.}$$

Thus, the final expressions have the form

$$(weight \times height)_{total} - (weight \times height)_{mec.} = heat_{mag-elec.} - heat_{voltaic\ elec.} \tag{2.5}$$

If

$$(weight \times height)_{total} - (weight \times height)_{mec.}$$

is positive, Joule says that the mechanical power has been converted into heat; if that value is negative, he says that heat has been converted into mechanical power. In all cases, the equation states

$$\alpha \textit{ units of mechanical power} = \beta \textit{ degrees of heat}.$$

Based on this, he calculates how many mechanical units correspond to one thermal unit. This is the *mechanical value of heat.*

Let us move on to the interpretation of the phenomena in relation to the heat theory. From the experiments which were performed in the first part of the paper, Joule concluded that heat can be created or destroyed. Therefore, heat cannot be a substance. If it is not a substance, it must be motion, according to the dilemma of that time. As heat is then a kind of motion and the calorific effects of the magneto-electricity are produced through motion, a motion causes another motion. A quantity of one kind is converted into the other. The conversion factor was experimentally determined. The thesis of the heat as motion was not new but Joule's novelty was significant (Appendix I).

2.2.2 1845: Gas Experiments and the Equivalent

In 1845, Joule published the article 'On the temperature changes produced by rarefaction and condensation'. This consists of 3 series of experiments that aim to show the correlation between *motion* and *heat*. In the first series, the amount of gas in a container is increased by mechanical action and the temperature of the gas rises. In the second, the gas leaves one of the containers and enters another, without external mechanical action. The temperature of both containers taken together remained constant. In the third, the gas leaves a container to perform mechanical action and the temperature of the gas goes down. Thus, we have:

- with condensation, heat increases;
- with rarefaction followed by condensation, heat is maintained;
- with rarefaction, heat decreases.

Now condensation and rarefaction are connected with mechanical actions. Therefore, increase and decrease of heat are correlated with mechanical action on the gas or by the gas. If there is condensation and rarefaction without mechanical action, the quantity of heat does not change. So mechanical action and heat appear correlated with each other.

Condensation experiments

The condensation experiments were performed with the mechanism shown in Fig. 2.10. A copper receiver is immersed in water. The air is compressed through the pump until approximately 22 atmospheres are reached.[93]

[93] "The pump and copper receiver were immersed in 45 lb. 3 oz. of water, into which the very sensible thermometer above described was then placed [...] The pump was then worked at a moderate degree

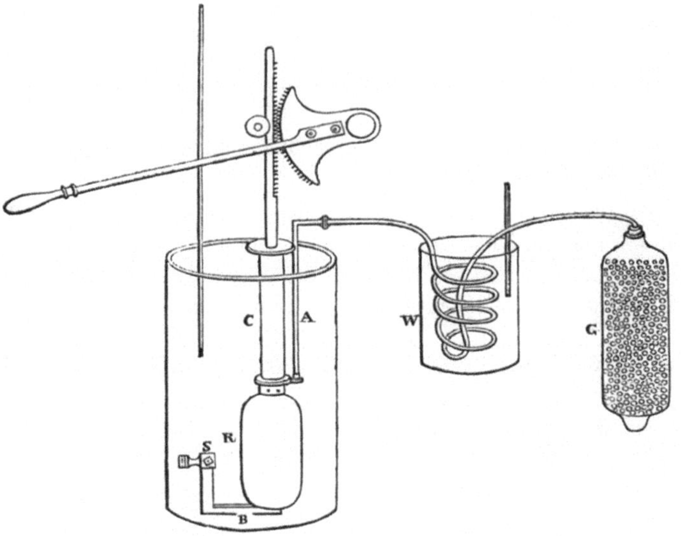

Fig. 2.10 A copper receiver (R) is immersed in water. The other container (G) contains pieces of calcium chloride intended to dry the air (Joule, 1845b, p. 173)

Calculating the force involved in condensing the gas and the heat developed, Joule puts them into the equation: to such a force corresponds such a heat, as he had done in the 1843 paper. He calculates then the mechanical equivalent of heat and arrives at the result: to raise by 1° F the temperature of a pound of water a force, capable of raising 823 lb. to the height of a foot, is required.[94] The second series of condensing experiments leads to a mechanical equivalent of heat equal to 795 lb.ft.[95]

Rarefaction Experiments

A certain amount of gas, which was at considerable pressure, is released. This gas is used to perform a motion, as shown in Fig. 2.11.

---

of speed until about twenty-two atmospheres of air, dried by being passed through the vessel G full of small pieces of chloride of calcium, were compressed into the copper receiver" (Joule 1845b, p. 176).

94 "a mechanical force capable of raising 823 lb. to the height of one foot must be applied in the condensation of air, in order to increase the temperature of a pound of water by one degree of Fahrenheit's scale" (ibid. p. 179).

95 "The mechanical force spent in the condensation is represented in this instance by [...] Hence the equivalent of a degree of heat per lb. of water, as determined by the above series, is 795 lb. raised to the height of one foot" (ibid. p. 180).

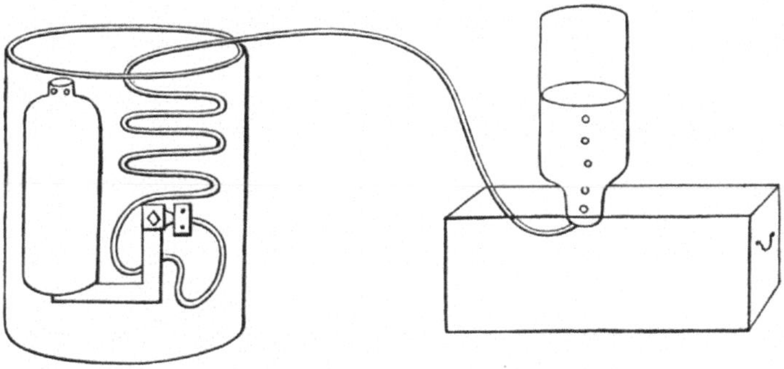

Fig. 2.11 A bottle with compressed air is immersed in a cylinder with water. When released, the gas is led into a container of water (ibid. p. 183).

When the stopcock is opened, the gas expands and is led into a container of water. The gas inlet forces the water out, which in turn raises the water level in the tub.[96] It 'pushes', therefore, a column of air at ambient pressure.

Joule determines the decrease in heat due to the expansion of the gas and calculates the work in moving the water as a result of the expanded air.[97] From the relationship between heat lost and work done a value for the mechanical equivalent of heat is obtained: 820 lb.ft.[98] The second and third series, which differ from each other by the quantities of air used, provide the values: 814 and 760 lb.ft.[99]

These three latter values are considered more reliable than the results obtained in the condensation experiments. Therefore, Joule takes the average of these,

$$\frac{820 + 814 + 760}{3} = 798\ \text{lb.ft}$$

[96] "The receiver was filled with dry compressed air, and a coiled leaden pipe, ¼ of an inch in internal diameter and 12 yards long, was screwed tightly upon the nozzle [...] The whole was then immersed into an oval can [...] and was also covered at top as perfectly as possible. Having ascertained the temperature of the water [...] the stopcock was opened and the air made to pass from the receiver through a pneumatic trough into a jar [...] After the air in the receiver had been reduced to the atmospherical pressure, the water was again well stirred and its temperature noted" (ibid. pp. 183–4).

[97] "The cold produced was diffused through 21.17 lb. of water, 14 lb. of copper, 8 lb. of lead, and 7 lb. of tinned iron. Hence we find that a quantity of cold was produced in the experiments sufficient to cause the temperature of a lb. of water to decrease by 4°.085. At the same time a mechanical force was developed, which could raise a column of the atmosphere, of an inch square at the base, to the altitude of 2723 inches; or, in other words, could raise 3352 lb. to the height of one foot" (ibid. pp. 184–5).

[98] "For each degree of heat lost there was therefore generated a force sufficient to raise 820 lb. to the height of one foot" (ibid. p. 185).

[99] "On reducing the results of these experiments in the manner before indicated, we find that in the experiments of Table V. 814 lb., and in those of Table VI. 760 lb. were raised to the height of a foot for every degree of heat per lb. of water lost" (ibid. p. 185).

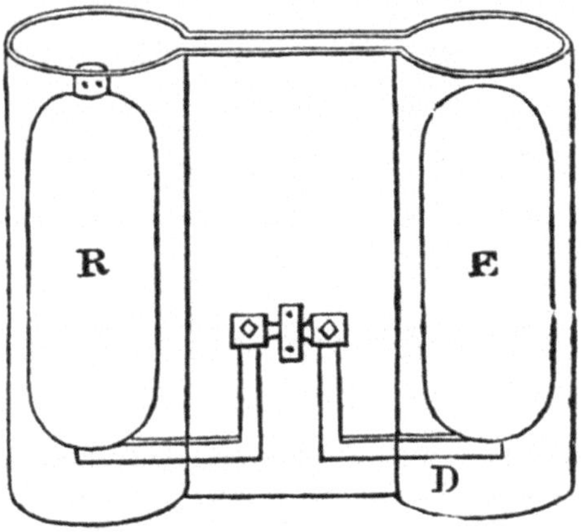

Fig. 2.12 Container R is filled with dry air at considerable pressure (about 22 atmospheres) and E has been exhausted (ibid. p. 180)

as the value for the mechanical equivalent of heat (4.29 J).[100]

The next experiments do not serve to provide values of the equivalent. They do, however, serve to complement the correlation between mechanical action and heat: when there is rarefaction and condensation without mechanical action, the quantity of heat in the system remains constant.

Rarefaction and condensation

The apparatus used (Fig. 2.12) consists of two copper receptacles connected together by a stopcock.[101] One is filled with dry air at considerable pressure (about 22 atmospheres) and the other has been exhausted. This set is immersed in a container of water, whose initial temperature is carefully measured. Opening the stopcock causes then the rarefaction in one container and condensation in the other. The temperature of the water is measured again.[102] The temperature variations are negligible in the various experiments. Joule considers the zero variation to be the correct value.[103]

100 "The mechanical equivalents of heat determined by the various series of experiments given in this paper are 823, 795, 820, 814, and 760. The mean of the last three, which I take as least liable to error, is 798 lb." (ibid. p. 187).

101 "I provided another copper receiver [...] which had a capacity of 134 cubic inches. Like the former receiver, to which it could be connected by a coupling nut, it had a piece D attached, in the centre of which there was a bore 1/8 of an inch diameter, which could be closed perfectly by means of a proper stop-cock" (ibid. p. 180).

102 "Having filled the receiver R [...] with about 22 atmospheres of dry air, and having exhausted the receiver E by means of an air-pump, I screwed them together, and then put them into a tin can containing 16½ lb. of water. The water was first thoroughly stirred, and its temperature taken by the same delicate thermometer [...] The stopcocks were then opened by means of a proper key, and the air allowed to pass from the full into the empty receiver until equilibrium was established between the two. Lastly, the water was again stirred and its temperature carefully noted" (ibid. pp. 181–2).

103 "we arrive at the conclusion, that *no change of temperature occurs when air is allowed to expand in such a manner as not to develop mechanical power*" (ibid. p. 182).

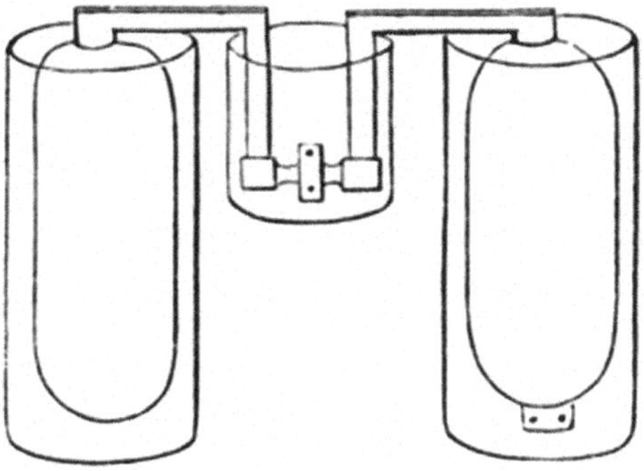

Fig. 2.13 Joule inverted the receivers and immersed them, as well as the connecting-piece, into separate cans of water (ibid. p. 183)

Joule then develops a strategy to determine the heat change in each container and around the stopcock. Each of the cylinders and the vicinity of the stopcock are immersed in disjoint recipients, as shown in Fig. 2.13.

In each of the water containers, the temperature of the water before and after the gas expansion is measured. The temperature has decreased in the container where R is immersed and increased in the other two. The sum of the variations in the containers where R and E are immersed is close to zero ($-2^\circ.36 + 2^\circ.38 = 0.02^\circ$).[104]

Conclusion

Joule establishes a correlation between heat and mechanical action based on the following:

- when mechanical action is done on the gas, heat is obtained
- when mechanical action is done by the gas, the gas loses heat
- when no mechanical action is involved, there is no heat gain or loss.

Based on the former two series of experiments, Joule determines the mechanical equivalent of heat. As in the 1843 paper, Joule shows the correlation between heat and mechanical action and determines the factor of conversion.

2.2.3 *1845–50: Paddle Wheel Experiments and the Equivalent*

In June 1845, at the meeting of the British Association for the Advancement of Science, Joule made known another way of determining the mechanical equivalent

[104] "One of the receivers had 2828 cubic inches of dry air condensed into it, while the other was vacuous. After equilibrium was restored by opening the cocks, I found that 2°.36 of cold per lb. of water had been produced in the receiver from which the air had expanded, while 2°.38 of heat had been produced in the other receiver, and 0°.31 of heat also in the can in which the connecting-piece was immersed, the sum of the whole amounting nearly to zero" (ibid. p. 183).

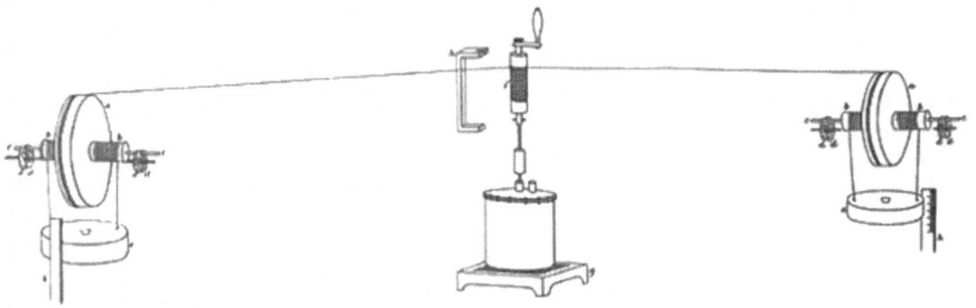

Fig. 2.14 Scheme of the setup

Fig. 2.15 Detail of the inside of the container

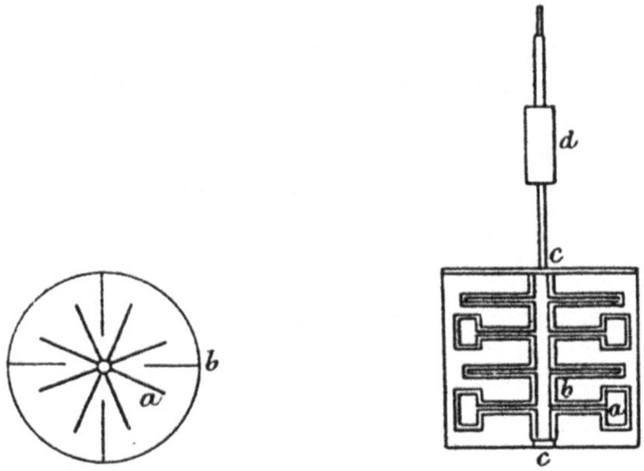

of heat. A short report of the experiment is given in the Association's report.[105] More detailed is the author's synopsis in a letter to the editors of the *Philosophical Magazine*, published by the journal that year (Joule, 1845a). At the 1847 meeting of the British Association in Oxford, Joule presented a paper on the mechanical equivalent of heat, whose experiments are analogous to those of 45.[106] In the 1850 article, published in an important journal of the time, the *Philosophical Transactions*, an illustration of the apparatus used is presented. I will use this illustration (Figs. 2.14, 2.15) to introduce the experiment.

The mechanism used consists of a paddle wheel immersed in water (other fluids were also used). When the axis of the paddle wheel is set in motion, the paddles propel the water. The container is, however, constructed in such a way as to allow

[105] British Association Rep. 1845, Trans. Chemical Sect. p. 31 (Joule, 1884, p. 202).

[106] "In the'Philosophical Magazine' for September 1845 I gave a concise account of some experiments [...] I have now repeated the experiments under more favourable circumstances, and with a more exact apparatus, and have moreover employed sperm-oil as well as water with equal success" (Joule, 1847a, pp. 277–8).

the paddles to rotate, but to hinder the rotational motion of the water. Some of the fluid hits plates solidly attached to the walls of the container.[107]

The paddle wheel is set in motion as follows. The axis of the paddle wheel is wrapped by a double wire, which is distributed between two diametrically situated pulleys. At the end of the wire passing through each pulley are suspended bodies, whose weight moves the paddle wheel.[108]

The experiment consists of dropping the bodies and measuring the temperature of the water before the bodies are dropped and at the end of the motion. The bodies motion occurs 16 times (1845) or 20 times (1847, 1850). As the wheel moves by virtue of the falling weights, Joule establishes a relationship between the mechanical power involved and the heat developed. The mechanical power is given by the weight and height of the bodies (as in 1843). The heat developed is calculated by means of the change in water temperature. Equating both mechanical power and heat developed, the mechanical equivalent of heat is calculated.

In 1845, the result was: 1°F, developed in water by friction, is equivalent to 890 pounds of the height of a foot.[109] In the 1847 experiment, Joule used water and whale oil. The mechanical equivalent of heat with water was 781.5 lb.ft and with oil 782.1 lb.ft. The average of these two values (781.8 lb.ft) was adopted as the mechanical equivalent of heat.[110] In an article in the *Comptes rendus*, Joule gives an account of the values of the mechanical equivalent of heat obtained in earlier experiments, with distilled water and whale oil. The results are presented in French units: 428.8 g at the height of 1 m is equivalent to the amount of heat needed to raise the temperature of 1 g of water by 1° centigrade. In the case of whale oil, 429.1 g

---

107 "The apparatus exhibited before the Association consisted of a brass paddle-wheel working *horizontally* in a can of water. Motion could be communicated to this paddle by means of weights, pulleys, & c." (Joule, 1845a, p. 203).

108 "The paddle moved with great resistance in the can of water, so that the weights (each of four pounds) descended at the slow rate of about one foot per second. The height of the pulleys from the ground was twelve yards, and consequently, when the weights had descended through that distance, they had to be wound up again in order to renew the motion of the paddle. After this operation had been repeated sixteen times, the increase of the temperature of the water was ascertained by means of a very sensible and accurate thermometer" (ibid. p. 203).

109 "A series of nine experiments was performed in the above manner [...] After reducing the result to the capacity for heat of a pound of water, it appeared that for each degree of heat evolved by the friction of water a mechanical power equal to that which can raise a weight of 890 lb. to the height of one foot had been expended" (ibid. p. 203).

110 "The equivalent of a degree of heat in a pound of water was therefore found to be 781.5 lb., raised to the height of one foot" (Joule, 1847a, p. 280). "the equivalent deduced from the friction of sperm-oil was 782.1, a result almost identical with that obtained from the friction of water" (ibid. p. 281). "The mean of the two results is 781.8, which is the equivalent I shall adopt until further and still more accurate experiments shall have been made" (ibid. p. 281).

lifted from 1 m.[111] Furthermore, he presents a result of a similar experiment, but performed with mercury. The result was 432.1 g at the height of 1 m.[112]

In the 1850 article, the frictions are performed with water, mercury and cast iron. From the experiments with water, the value of the mechanical equivalent heat was 773.64 lb.ft.[113] The experiments with mercury were carried out in two series. The first led to the value of 773,762 and the second, performed with fewer weights, led to 776,303 lb.ft.[114] From the experiments with cast iron, a value of 776.997 was obtained and in the second series, with less weights, 774.88 lb.ft.[115]The experiments with water are considered the most reliable. Hence the value of the equivalent *in vacuo* for the friction of water, 772.692, is adopted as the mechanical equivalent of heat.[116]

Joule ended this paper with three propositions but the third one was suppressed in accordance with the wish of the Committee who reviewed the paper. The published conclusions are the following:

> "1st. That the quantity of heat produced by the friction of bodies, whether solid or liquid, is always proportional to the quantity of force expended. And,
>
> 2nd. That the quantity of heat capable of increasing the temperature of a pound of water (weighed in vacuo, and taken at between 55° and 60°) by 1°F requires for its evolution the expenditure of a mechanical force represented by the fall of 772 lb. through the space of one foot" (ibid. p. 328).

---

[111] "Quand l'eau était agitée par l'action d'une roue à pannes agissant dans le liquide [...] La force mécanique capable d'élever un poids de 428.8 grammes à la hauteur de 1 mètre fut ainsi trouvée être l'équivalent d'une quantité de chaleur nécessaire pour élever la température de 1 gramme d'eau par 1 degré centrigrade. J'ai aussi fait des expériences semblables sur la friction de l'*huile de* baleine [...] le développement de 1 degré de chaleur par gramme d'eau était égal à 429.1 grammes soulevé de 1 mètre" (Joule, 1847b, pp. 283–4).

[112] "Poursuivant mes recherches, j'ai aussi employé du *mercure* comme liquide frotté" (ibid. p. 284). "Par conséquent, la chaleur capable d'augmenter la température de 1 gramme d'eau de 1 degré centrigrade est égale à une force mécanique capable d'élever un poids de 432.1 grammes à 1 mètre de hauteur" (ibid. pp. 285–6).

[113] "773.64 foot-pounds will be the force which, according to the above experiments on the friction of water, is equivalent to 1° Fahr. in a lb. of water" (Joule 1850, p. 312).

[114] "773.762; which is therefore the equivalent derived from the above experiments on the friction of mercury" (ibid. p. 318). "776.303 will therefore be the equivalent from the above series of experiments, in which the amount of friction of the mercury was moderated by the use of lighter weights" (ibid. p. 321).

[115] "776.997 will therefore be the equivalent derived from the above experiments on the friction of cast iron. The next series of experiments was made with the same apparatus, using lighter weights" (ibid. p. 325). "774.88 will therefore be the equivalent as derived from this last series of experiments" (ibid. p. 327).

[116] "I consider that 772.692, the equivalent derived from the friction of water, is the most correct, both on account of the number of experiments tried and the great capacity of the apparatus for heat" (ibid. p. 328).

Conclusion

At the beginning of a paddle wheel experiment, the bodies are at a certain height and the water is a certain temperature. At the end, the bodies are down and the temperature of the water has increased. Thanks to these data the following relationship is established

$$weight \times height = \beta \; units\, of\, heat.$$

If α units of mechanical power correspond to β units of heat, to one unit of heat corresponds x, the value of the mechanical equivalent of heat.

All this leads to the two propositions of Joule's conclusion quoted above. The third proposition, which was not published, stated: 'friction consisted in the conversion of mechanical power into heat'.[117] Indeed, this statement cannot be proved by those experiments. The process of conversion, if it takes place, must have happened within the can. However, no observation is made to verify what happened within it. The temperature of water is read off when the weights are down and there is no conversion anymore, if there were any. In fact, what really happened within the can is also not important for the final result: no specific information of the process of conversion is used. If the phenomenon does not consist of a process of conversion, the established relationship between heat and mechanical power is not disturbed. 'Conversion' is therefore an interpretation of the phenomenon. Thus, it is understandable that Faraday, who reviewed the paper (Smith, 1976), did not accept the third conclusion.

2.2.4 Conclusion

In all of Joule's articles (1843–50) there are two important topics: the mechanical equivalent of heat and the thesis 'heat is motion'. Joule designed and performed experiments by means of which he was able to measure the mechanical power and heat evolved. He established the numerical relation between both magnitudes and determined the mechanical equivalent of heat. In the course of time, he understood that the values of the equivalent could be more precise than in 1843 and 1845. The aim of the 1850 paper was to determine more accurately the mechanical equivalent of heat.[118] (Fig. 2.16).

The other recurrent topic in Joule's articles concerns the theory of heat. Contrary to the thesis in use at that time, that heat is a substance, Joule argues that heat is

[117] "A third proposition, suppressed in accordance with the wish of the Committee to whom the paper was referred, stated that friction consisted in the conversion of mechanical power into heat" (ibid. p. 328).

[118] "[...] left no doubt on my mind as to the existence of an equivalent relation between force and heat; but still it appeared of the highest importance to obtain that relation with still greater accuracy. This I have attempted in the present paper" (ibid. p. 302).

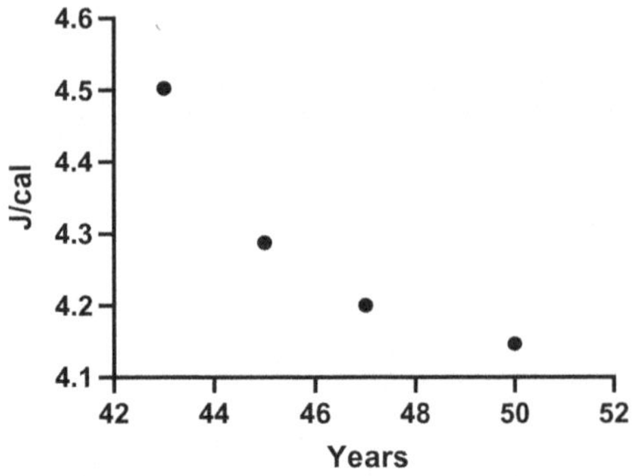

Fig. 2.16 1843, 45, 47 mean values and the 1850 adopted one

motion. This thesis appeared in 1843 and is maintained in all other papers. The 1850 article is symptomatic in this regard. This writing opens with a quotation of Locke: "Heat is [...] *motion*" and another of Leibniz: "The *force* of a moving body is proportional to the square of its velocity".[119] Rumford is quoted because he said, heat is motion.[120] Davy's experiment of rubbing two pieces of ice together is also referred to.[121] Joule's thesis was not new, but his contribution to its acceptance—the mechanical equivalent of heat—was decisive.

2.2.5 *Mayer and Joule*

Mayer and Joule knew that at the time the thesis 'heat is substance' was valid. This thesis had experimental support. Based on experiments conducted over the years, most of the authors argued that the amount of heat did not vary.[122]

119 ""Heat is a very brisk agitation of the insensible parts of the object, which produces in us that sensation from whence we denominate the object hot; so what in our sensation is *heat*, in the object is nothing but *motion*."—Locke.

"The *force* of a moving body is proportional to the square of its velocity, or to the height to which it would rise against gravity."—Leibnitz"" (ibid. p. 298).

120 "For a long time it had been a favourite hypothesis that heat consists of "a force or power belonging to bodies", but it was reserved for Count Rumford to make the first experiments decidedly in favour of that view [...] "It appears to me," he remarks, "extremely difficult, if not quite impossible, to form any distinct idea of anything capable of being excited and communicated in the manner the heat was excited and communicated in these experiments, except it be motion"" (ibid. pp. 298–9).

121 "By rubbing two pieces of ice against one another in the vacuum of an air-pump [...] This experiment was the more decisively in favour of the doctrine of the immateriality of heat, inasmuch as the capacity of ice for heat is much less than that of water. It was therefore with good reason that Davy drew the inference that "the immediate cause of the phenomena of heat is motion, and the laws of its communication are precisely the same as the laws of the communication of motion"" (ibid. p. 300).

122 There were two main theses concerning the nature of heat. According to Rumford (1798, p. 99) or Davy (1799, p. 13–4), heat was motion. According to Carnot (1824, p. 10–1, 28) or William

Experimental	*Conceptual*
The quantity of heat does not vary	Heat is a substance

Both Mayer and Joule show that motion gives rise to heat (increasing in quantity) and heat gives rise to motion (decreasing in quantity). Since the quantity of heat varies after all, "heat is a substance" is no longer valid. With the collapse of this thesis, there is room for another thesis on the nature of heat, that is, the question arises, what is heat then? Mayer and Joule forge their own concepts of heat. In both concepts, heat and motion appear related. For Mayer, they are both forces. Motion and heat are forms of force. For Joule, they are both motions. Heat is a kind of motion.

Experimental	Conceptual
The quantity of heat does not vary	Heat is substance
The quantity of heat varies	Heat is a form of force Heat is motion

The terms that each of them uses for phenomena involving heat and motion are consistent with these concepts. Mayer uses *transformation.* This term tells us that there is a change of form. Now, since heat and motion are forms of force, the term transformation is consistent in cases where one gives rise to the other. Joule uses *conversion.* Now since heat is a type of motion, if motion gives rise to heat or inversely, we have one type of motion converting to another.

Joule's thesis on heat fits the dilemma of the time: whether heat should be a substance or motion. Therefore, showing that heat could not be a substance, it remained to be a motion. Mayer's thesis does not fit. He stressed in his very first writing that he did not argue that heat was motion. In the last book, he denies that there was any knowledge about the nature of heat. For Mayer, heat is one of the five forms of force.

The mechanical equivalent of heat is the especially important point of convergence. Mayer determined it in 1842 and 1845. Joule determined this in 1843 and in the following articles based on experiments that he himself designed and carried out (Fig. 2.17).

2.3 Ludwig Colding

Ludwig Colding was an engineer, newly graduated, unemployed, who got a grant to do a research work. He took this opportunity to prove an idea he had had for a long time and could not get rid of *the forces of nature do not perish.* Now, by observation,

---

Thomson (1849, p. 315), heat was a substance. Some authors had posed the question, "what is heat?", in connection with their experimental works during the first part of the nineteenth century. This was the case of Haldat (1807, p. 214), Berthollet (1809, p. 447), in cooperation with Pictet and Biot, Colladon and Sturm (1828, p. 161). Joule's research concerns this question: heat is either a substance or motion (see Appendix A).

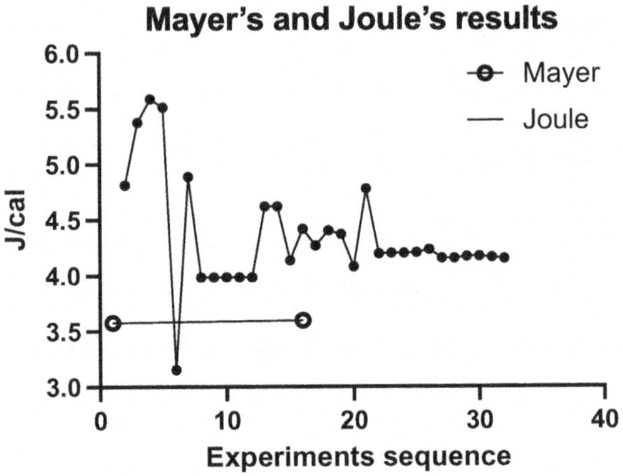

Fig. 2.17 All the values of the mechanical equivalent of heat obtained by Mayer and Joule between 1842 and 1850 and expressed in modern units (included is the value of 770 lb.ft/BTU that appears in the 1843 PS to the paper (Joule, 1843, p. 442) and had not yet been mentioned.)

we realize that if a force acts on a particle and causes a movement, this propagates to the surrounding environment eventually becoming imperceptible.[123] But this is no reason, Colding continues, to say that something has been lost without any effect. Rather, he thinks that it is in the very nature of things, that the forces we perceive as having disappeared, act in a different way.[124]

Colding tries to make this idea a general law of nature.[125] To achieve this goal, he does not do research on various forces but focuses on the disappearance of mechanical force and the appearance of heat. The result of his research was presented to the Danish Society of Sciences in 1843. The paper *Theses on force* was published in 1856.

2.3.1 1843: Force, Heat and the Proportionality

In 1843, Colding defends the thesis that forces of nature are imperishable. He begins the paper by mentioning several studies that corroborate his thesis. These

123 "When certain moving forces act on a material particle […] a quantity of motion is created, proportional to the acting force. This quantity of motion, in turn, is transmitted to the surrounding material particles and propagated from them in the same manner, without cessation, so that within a short time the originally introduced quantity of motion is distributed through such a large mass that no perceptible trace of this activity remains" (Colding 1856, p. 1).

124 "it does not seem to me that there is any justification for assuming that some activity may be gradually lost in matter without in any way appearing as perceptible effect in its original amount; it seems to me even more in the nature of things that those forces which seemingly vanish must again appear, acting in other ways. This thought occurred to me long ago, and I have never been able to discard it" (ibid. p. 1).

125 "I have […] become so convinced of the validity of this thesis that I will attempt to propose as a general law of nature: when a force seems to disappear it merely undergoes a transformation, whereupon it becomes effective in other forms" (ibid. p. 1).

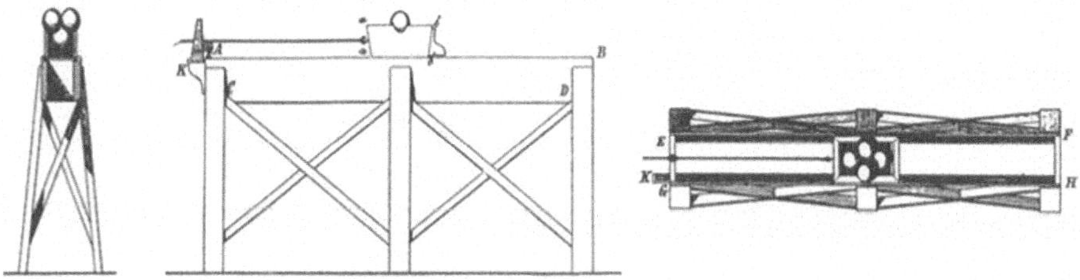

Fig. 2.18 Colding's apparatus from the front, side and top viewpoints. The apparatus consists of two parallel brass bars, a little over two meters long, on which a small "sled" slides (Colding, 1856)[127]

are compression experiments (with gases, liquids and solids), friction experiments and experiments on breaking iron rods (Colding, 1856, p. 1–2) The results of these studies, however, are not sufficient to prove his thesis. Hence, he undertakes some laboratory work. The experiments carried out consist of moving a small sled over bars and determining the variation of heat as a function of the weight of the sled.[126] This mechanism, represented in Fig. 2.18, is analogous to Coulomb's in the study of mechanical machines.

The sled carries balls, the number of which allows its weight to vary.[128] It is pulled by hand thanks to a string attached to it at one end.[129] The distance covered by the sled is the same in all experiments. The course is performed twice with roughly the same speed.[130] The expansion of the ribbons at the bottom of the sled and the bars of the rail provide an indication of the heat developed. An instrument, which reacts to the variation of length and gives an account of the variation of heat, is attached to the sled and another one to one of the bars.[131]

126 "However, since the results which emerge from the indicated experiments are quite insufficient for deciding whether the proposed law is correct or not, and since to my knowledge other significant experiments on the evolution of heat from friction are not available for deciding this, I have myself undertaken [...] a series of experiments with various solid bodies subjected to different pressures and speeds. Still, these experiments should only be regarded as preliminary [...] since the apparatus which I have utilized is not sufficiently accurate" (ibid. p. 3).

127 On Colding's apparatus, although constructed about 1844, see Kragh (2009, p. 9).

128 "The pressure is produced by a sled-load [...] and the force necessary to pull the sled is obtained from a dynamometer" (ibid. p. 4).

129 "The sled is pulled by hand with the aid of a rope and is guided by lists of wood along the sides of the track" (ibid. p. 4).

130 "The speed with which the track was traversed was approximately 2 feet per second in all measurements" (ibid. p. 4).

131 "Consider an apparatus similar to that used by *Coulomb* in his investigation of friction; namely, a sled whose runners are fitted with underlying metal strips which are to be examined in regard to the heat produced by friction. The track on which it slides is overlaid with a similar pair of rails. If the sled, loaded with cannon balls, is moved from one end of the track to the other, these surfaces will expand because of the heat from friction, and if the expansion of the sled runner as well as the track is measured with a delicate sensing lever, one can read off the quantity of heat evolved by the friction, since the quantity of heat [...] can be taken to be proportional to the degree of heat, or expansion [...]" (ibid. p. 3).

Table 2.1 The mean values of the first three series

Moving force	11	19.7	30.3
Heat	0.72	1.32	2

Table 2.2 The normalized mean values of the series

Moving force	1	1.79	2.75
Heat	1	1.83	2.77

Table 2.3 The normalized mean values of the seven series

Force	1	1.24	1.68	1.74	1.77	1.79	2.75
Heat	1	1.20	1.66	1.80	1.76	1.83	2.77

Colding performs 10 series of experiments. The difference between them is the weight of the sled and the materials used for the ribbons: brass, zinc, lead, linden wood, linden wood wrapped in flannel and iron (ibid. p. 6–11). In all the series, the driving force employed to pull the sled is measured thanks to a dynamometer, and the heat developed, through his sensors. Let us consider the first three series.

In the first, the weight of the sled was 88.75 lb. (44.375 kg). Thirty-two measurements were taken, as a result of which the average value of the force and the heat developed was calculated: 30.3 lb. for force and 2° of heat.

In the second series, the sled weighed 53.5 lb. (26.75 kg). Fifty measurements were taken and the average values were: 19.7 lb. for force and 1.32° of heat.

In the third series, the weight of the sled was 31 pounds. Fifty-six measurements were taken, with the results: 11 pounds for force and 0.72° of heat (Table 2.1).

Colding improves the presentation of the results by expressing the values of force and heat through the latter values, 11 and 0.72 respectively. Thus, the ratio between 30.3 lb. of the first series and 11 lb. is 2.75; and the ratio between 2° of heat and 0.72 is 2.77. Analogously, for the second series:

$$\frac{19.7}{11} = 1.79 \text{ and } \frac{1.32}{0.72} = 1.83.$$

Putting the data together, we have (Table 2.2).

The results of the remaining series are presented in the same way. Displayed in ascending order, we have (Table 2.3, Fig. 2.19)[132]:

Given these values, Colding concludes that the amount of heat developed is proportional to the motive force lost.[133] This result, together with the studies on the relation of lost force and gained heat referred to in the beginning, satisfactorily confirm

[132] Besides these results, there are three more, which the author does not consider valuable.

[133] "From these two series of ratios it is clear that we are justified in concluding that the quantities of heat evolved are in every case proportional to the lost moving forces" (ibid. p. 12).

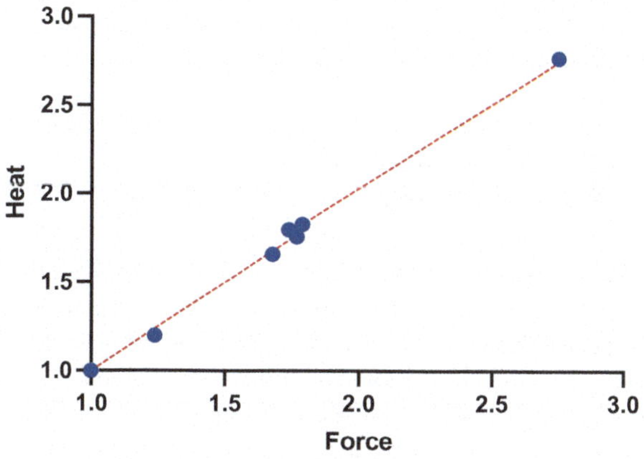

Fig. 2.19 The mean values obtained by Colding and the dashed line *force* ≈ *heat*

his thesis, says Colding, 'when a force seems to disappear it merely undergoes a transformation and reappears in other forms' (ibid. p. 13).

2.3.2 *Conclusion*

Colding assumes that the forces of nature do not perish. Therefore, what disappears reappears in another form. Experiments are conducted to corroborate this idea. Observation shows that forces disappear, but Colding argues, they do not destroy themselves but are transformed. The elements of this transformation are observable, such as mechanical action and heat. A way of equating this relationship is proposed: if the mechanical action is quantitatively representable by q, the effect must be set equal to q. Using this expression, we would arrive at an equation of the form

$$q(mechanical\ units) = Q(units\ of\ heat).$$

Colding did not write an equation like this, which could have provided him with a value of the mechanical equivalent of heat. What he concludes from his experimental work is that

variation of force is proportional to the variation of heat

$$\Delta \mathrm{F} \approx \Delta \mathrm{H}.$$

Taking into account the thesis of imperishability of the forces of nature and this result, one realizes that the contribution of the result to the thesis lies in the following. The more force one applies, the more heat appears. Therefore, the increase in force was not lost; it appeared as heat. If Colding had employed more force and the heat developed was the same, he would have no argument to defend imperishability. His experimental work satisfies, therefore, a necessary condition for the validity of the imperishability of the forces of nature.

2.3.2.1 Mayer and Colding

Mayer advocates the indestructibility of force. Colding advocates the imperishability of the forces of nature. Mayer's indestructibility is based on the indestructibility of causes, and this, in turn, is based on the equality of cause and effect. Colding's imperishability is a conviction: it is an idea he has had for some time. Whether Mayer's causal relationship holds for a given phenomenon or not is decided by means of observation. To use Colding's imperishability with regard to a phenomenon, we need to know what a force of nature is, which is not defined. Both Mayer and Colding use the term transformation to justify what is different but correlated in a phenomenon, such as heat and motion.

Mayer determined the mechanical equivalent of heat. Colding presents a proportionality between motive force and heat. Mayer's force has been translated by energy. Colding's force cannot be. This is the force that is measured by a dynamometer.[134]

2.3.2.2 Joule and Colding

Joule determined the mechanical value of heat in 1843. Colding performed a series of experiments but did not determine it, even though he had theoretically operationalized the relationship of magnitudes that could have led to the mechanical equivalent of heat. Joule interpreted the result as a consequence of the indestructibility of natural powers.[135] This is analogous to Colding's imperishability of the forces of nature. Colding's imperishability played a role in his research. Joule came up with the indestructibility of natural powers after the research had been done.

2.4 Hermann von Helmholtz

In 1847, Hermann von Helmholtz was a practicing physician. That year he presented a paper to the Physical Society of Berlin entitled *On the conservation of force*, which was published in the same year at his own expense. In this short book, he argues that there are two ultimate forces in nature, whose sum remains constant. (This is the sense of *conservation of force* in the title of the book.) One of these forces is called tension force and the other living force. Tension force is for example a body at a

[134] Based on the data presented in Table 2.3, Dahl calculated a value for the mechanical equivalent of heat (Colding, 1972, p. 178–9). See also Kragh (2009, p. 9).

[135] "the grand agents of nature are, by the Creator's fiat, *indestructible*; and that wherever mechanical force is expended, an exact equivalent of heat is *always* obtained." (Joule, 1843, p. 158).

certain height above the ground; the falling motion is living force. A body at a certain height that is dropped is not only an example. It is the model of the conception. Heat, electrical and electromagnetic phenomena are conceived in the same way: as if there were particles, which acted on by tension forces go into motion. Of course, in thermal phenomena or electrical circuits, we do not observe such forces. In such cases, they can be imagined. In fact, Helmholtz needs no more than this possibility of imaging motion and tension forces, because when it comes to dealing with a phenomenon, he tells us what is to be taken as a tension force and living force in that phenomenon. Since we thus know what corresponds to both forces, the motion of the particles is superfluous. In a second step, Helmholtz puts the forces in an equation. So, if one increases, the other decreases in the same proportion. To support this, he presents two arguments. First, these forces have a reciprocal causal relationship; second, the motive force and the resultant motion are equivalent. This latter proposition, which is taken as a principle, is drawn from the idea, ascribed to Carnot and Clapeyron[136]: it is impossible to get a lasting motive force out of nothing. Let us see how this works in phenomena.

2.4.1 1847: Conservation of Ultimate Forces

Free fall is the first phenomenon in analysis. Based on the principle just mentioned, the free fall motion (which is taken as a force) requires a motive force. This motive force is taken equivalent to the weight and height at which the body was at the beginning of the fall. The motive force is called force of tension and the other, *vis viva* (living force). By living force is not designated the magnitude mv^2, as we saw before, but half of it.[137] Thus, the equality between the forces comes in the form,

$$\frac{1}{2}mv^2 = weight \cdot height \tag{2.6}$$

Weight is given by the mass of the body and acceleration due to gravitation,[138]

$$\frac{1}{2}mv^2 = mgh \tag{2.7}$$

---

[136] "Wir gehen aus von der Annahme, dass es unmöglich sei durch irgend eine Combination von Naturkörpern bewegende Kraft fortdauernd aus nichts zu erschaffen. Aus diesem Satze haben schon *Carnot* und *Clapeyron* eine Reihe [...]" (Helmholtz 1882, p. 17).

[137] "Der besseren Uebereinstimmung wegen mit der jetzt gebräuchlichen Art die Intensität der Kräfte zu messen, schlage ich vor, gleich die Grösse 1/2 mv^2 als Quantität der lebendigen Kraft zu bezeichnen, wodurch sie identisch wird mit dem Maasse der Arbeitsgrösse" (ibid. p. 18).

[138] "Die Arbeitsgrösse, welche gewonnen und verbraucht wird, kann bekanntlich ausgedrückt werden als ein auf eine bestimmte Höhe h gehobenes Gewicht m; sie ist dann $m\ g\ h$, wo g die Intensität der Schwerkraft. Um senkrecht frei in die Höhe h emporzusteigen braucht der Körper m die Geschwindigkeit $v = \sqrt{2gh}$; und erlangt dieselbe wieder beim Herabfallen. Es ist also $½mv^2 = mgh$" (ibid. p. 18).

The next step consists in generalizing the principle to central forces.[139] The variation of the living force of a body or atom, more exactly, the difference between the living force at the end of the path and at the beginning, is attributed to the acting forces. Thus, we have

$$\textit{Living force}(\text{at the end}) - \textit{Living force}(\text{at the beginning}) = \textit{force of tension},$$

$$\frac{1}{2}mv^2_{(final)} - \frac{1}{2}mv^2_{(initial)} = -\int_r^R \varphi dr \tag{2.8}$$

where φ stands for the acting force and r for the distance between the bodies.[140] The equation is then read as follows: the increase in the living force of a moving point under the influence of a central force is equal to the sum of the tension forces.[141] Thus, the sum of the living and tension forces is constant, which is the content of the principle of conservation of force.[142]

---

139 "Wir wollen dem besprochenen Gesetze für die Fälle, wo Centralkräfte wirken, nun noch einen allgemeineren Ausdruck geben" (ibid. p. 21).

140 The minus sign of the integral is conventional: the intensity of the force is taken as positive in attraction and negative in repulsion. If the final velocity is greater than the initial velocity then the difference between the two is positive. The final distance R is smaller than the initial r because the moving body was farther away from the body that is attracting it and has moved closer. If the primitive of φ(R) is less than that of φ(r), then the minus sign before the integral becomes necessary. If the primitive of φ(R) is greater than that of φ(r), the minus sign is not necessary, but the tension force decreases with the distance. Helmholtz took the former option. "Ist φ die Intensität der Kraft, welche in der Richtung von *r* wirkt, wenn sie anzieht, als positiv, wenn sie abstösst, als negativ gesetzt [...] wenn *Q* und *R*, *q* und *r* zusammengehörige Tangentialgeschwindigkeiten und Entfernungen vorstellen:

$\frac{1}{2}mQ^2 - \frac{1}{2}mq^2 = -\int_r^R \varphi dr$ (ibid. pp. 21–2).

141 "Um die Bedeutung der Grösse $\int_r^R \varphi dr$ zu finden, denken wir uns die Intensitäten von φ, welche zu verschiedenen Punkten der Verbindungslinie von *m* und *a* gehören [...] so ist jene Grösse der Inbegriff aller Kraftintensitäten, welche in den zwischen *R* und *r* liegenden Entfernungen wirken. Nennen wir nun die Kräfte, welche den Punkt *m* zu bewegen streben, so lange sie eben noch nicht Bewegung bewirkt haben, *Spannkräfte*, im Gegensatz zu dem, was die Mechanik *lebendige Kraft* nennt, so würden wir die Grösse $\int_r^R \varphi dr$ als die *Summe der Spannkräfte* zwischen Entfernungen *R* und *r* bezeichnen können" (ibid. p. 22).

142 "In allen Fällen der Bewegung freier materieller Punkte unter dem Einfluss ihrer anziehenden und abstossenden Kräfte, deren Intensitäten nur von der Entfernung abhängig sind, ist der Verlust an Quantität der Spannkraft stets gleich dem Gewinn an lebendiger Kraft, und der Gewinn der ersteren dem Verlust der letzteren. *Es ist also stets die Summe der vorhandenen lebendigen und Spannkräfte constant.* In dieser allgemeinsten Form können wir unser Gesetz als das *Princip von der Erhaltung der Kraft* bezeichnen" (ibid. pp. 24–5).

The presentation of this principle in Mechanics[143] is followed by the defense of its validity for the other domains of Physics. The sections of the book concerning these domains are significantly titled: "force equivalent of heat", "force equivalent of electrical phenomena" and "force equivalent of magnetism and electromagnetism". The final part of the book, on living beings, has no title.

The force equivalent of heat

The term 'force equivalent of heat' could lead one to think of the quantity of force equivalent to a given amount of heat. If that quantity were the thermal unit, then it would be the mechanical equivalent of heat. This is not the case. 'Equivalent force of heat' refers to the way force is viewed in heat phenomena, i.e., what is to be taken as tension force and as living force in these phenomena. This point of view presupposes that heat is motion, which contradicted most physicists of that time for whom heat was a substance. Helmholtz will understandably counter-argue the thesis of heat as a substance. He will also argue against this thesis that force is not conserved. Let us start with these counterarguments.

Helmholtz addresses the cases in which absolute loss of force was admitted: inelastic collision and friction.[144] His argument goes as follows. In the collided body, there is a change in shape and a change in density, thus an increase in the force of tension. Furthermore, when the collision is repeated, as for example when a metal is hit with a hammer, heat is detected. On the other hand, some of the effect passes into the air as sound.[145] In friction, besides the change of shape and the displacement of particles on the surface, there are thermal and electrical changes.[146] In short, to the absolute loss of force in inelastic collision and friction, Helmholtz counterposes effects arising from these phenomena. He is not interested in the various effects, but in the heat effect.

He then asks the question of whether in a loss of mechanical force a certain amount of heat arises. The answer to this is in the affirmative. The argument put forward comes from Joule's water friction experiments from 1843 and 1845.[147] (The value of the mechanical equivalent of heat as determined by Joule is called into

---

[143] "Wir gehen jetzt zu den speciellen Anwendungen des Gesetzes von der Constanz der Kraft über. Zuerst haben wir diejenigen Fälle kurz zu erwähnen, in denen das Princip von der Erhaltung der lebendigen Kraft bisher schon benutzt und anerkannt ist" (ibid. p. 27).

[144] "Diejenigen mechanischen Vorgänge, bei welchen man bisher einen absoluten Verlust von Kraft angenommen hat, sind: (1) *Der Stoss unelastischer Körper* [...], (2) *Die Reibung*" (ibid. pp. 31–2).

[145] "Derselbe ist meist mit einer Formveränderung und Verdichtung der gestossenen Körper verbunden, also mit Vermehrung der Spannkräfte; dann finden wir bei oft wiederholten Stössen der Art eine beträchtliche Wärmeentwickelung, z. B. beim Hämmern eines Metallstückes; endlich wird ein Theil der Bewegung als Schall an die anstossenden festen und luftförmigen Körper abgegeben" (ibid. pp. 31–2).

[146] "sowohl an den Oberflächen zweier sich über einander hinbewegender Körper, als im Innern derselben bei Formveränderungen [...] Auch bei der Reibung finden meistens geringe Veränderungen in der moleculären Constitution der Körper [...] Ausserdem finden aber stets auch thermische und electrische Aenderungen statt" (ibid. p. 32).

[147] "Zur Lösung der ersteren Frage sind erst wenige Versuche angestellt. *Joule* hat die Wärmemengen untersucht, welche bei der Reibung des Wassers in engen Röhren und in einem

question, but the existence of an equivalence between mechanical power and heat is admitted.[148])

He then asks the question: to what extent can an amount of heat correspond to a mechanical force? There is no direct answer to this question. Helmholtz goes on to attack the heat substance thesis instead. The theory of heat as a substance must necessarily admit that the amount of heat remains constant. In that case, the production of work must originate in the expansion of the substance, says Helmholtz.[149] This thesis is not directly denied. He rather goes on to show that the heat-substance concept is not adequate for the explanation of thermal phenomena in general. The explanation of heat arising from friction is the first difficulty pointed out in that theory. The heat-substance theory would have to admit that the heat comes from outside.[150] This explanation has, however, no experimental proof, Helmholtz argues.[151]

An alternative to this explanation, due to Berthollet, is to admit that the heat of friction is originated by the compression of the surface and the rubbed parts. Helmholtz counterargues, this explanation is contradicted by several experiments, such as the melting of ice by friction performed by Davy.[152]

Another counterargument looks at the material theory of heat from another point of view: the amount of heat can absolutely be increased through motion. (If it can be increased, then the amount of heat varies. Therefore, heat cannot be substance). Here the experiment with the magneto-electric machine (Joule, 1843) is adduced. As long as there is motion of the magnet in the vicinity of the coil, there is electric current, which gives rise to heat.[153] Now, continues Helmholtz, heat cannot come from the parts of the machine, since it can be produced indefinitely.[154] Therefore, concludes

---

Gefässe entwickelt werden, wo es durch ein nach Art einer Turbine construirtes Rad in Bewegung gesetzt wurde" (ibid. p. 33).

148 "Indessen entsprechen seine Messungsmethoden zu wenig der Schwierigkeit der Untersuchung, als dass diese Resultate irgendwie auf Genauigkeit Anspruch machen könnten; wahrscheinlich sind diese Zahlen zu hoch, weil [...]" (ibid. p. 33).

149 "Die materielle Theorie der Wärme muss nothwendig die Quantität des Wärmestoffs als constant ansehen; mechanische Kraft kann er nach ihr nur durch sein Streben sich auszudehnen erzeugen. Für sie kann das Kräftäquivalent der Wärme also auch nur in der Arbeit bestehen, welche dieselbe bei ihrem Uebergang aus einer höheren in eine niedere Temperatur leistet; in diesem Sinne haben *Carnot* und *Clapeyron* die Aufgabe bearbeitet [...]" (ibid. p. 33).

150 "Um die Reibungswärme zu erklären, muss die materielle Theorie entweder annehmen, dass dieselbe von aussen zugeleitet sei, nach *W. Henry*, oder dass dieselbe nach *Berthollet* durch Compression der Oberflächen und der abgeriebenen Theile entstehe" (ibid. pp. 33–4).

151 "Der ersteren Annahme fehlt bisher noch jede Erfahrung" (ibid. p. 34).

152 "die zweite [...] scheitert ganz bei der Reibung von Flüssigkeiten und bei den Versuchen, wo Eisenkeile durch Hämmern glühend und weich gemacht, Eisstücke durch Reibung geschmolzen werden" (ibid. p. 34).

153 "Derselbe Fall findet bei den magnetelectrischen Maschinen statt; so lange Magnet und Anker gegeneinander bewegt werden, entstehen electrische Ströme, welche im Schliessungsdraht Wärme erzeugen" (ibid. p. 35).

154 "Es kann hier offenbar aus den die Maschine constituirenden Körpern in das Unendliche Wärme entwickelt werden, ohne dass dieselbe irgendwo verschwände" (ibid. p. 35).

Helmholtz, heat phenomena cannot be explained by a substance, to which he adds, they must be explained as motion.[155]

About the motion of which heat would consist, an idea is given, but announced as a mere hypothesis.[156] Indeed, it is not necessary to specify the kind of motion that heat consists of. It is enough for Helmholtz that heat phenomena can be thought of as motion.[157] Let heat then be a kind of motion. How can the ultimate forces of nature, tension forces and living forces be understood then? The quantity of heat, says Helmholtz, is to be understood as the sum of the living forces of heat motion and the forces of tension in the atoms. (Atoms were not observable. Therefore, there were no experimental conditions to test the thesis.) The heat connected to the living forces would correspond to the so-called free heat (the heat that is detectable by a thermometer) and the heat connected to the tension forces, to the latent heat.[158]

Finally, Helmholtz highlights the explanatory power of the theory. The conservation of force could account for the phenomena of heat conduction and radiation, among others, that until then had been explained by the conservation of the quantity of heat-substance.[159] The theory would also explain the phenomena of the appearance of heat in chemical processes. The argument is: the forces of chemical attraction (tension forces) would give rise to motion, whose amount of living force would manifest itself in the heat arising.[160] The theory would also be able to subsume the disappearance of heat where there is the production of work. Here Joule's (1845a, b) experiment, in

---

[155] "Aus diesen Tathsachen folgt nun, dass die Quantität der Wärme absolut vermehrt werden könne durch mechanische Kräfte, dass deshalb die Wärmeerscheinungen nicht hergeleitet werden können von einem Stoffe, welcher durch sein blosses Vorhandensein dieselben bedinge, sondern dass sie abzuleiten seien von Veränderungen, von Bewegungen" (ibid. p. 35).

[156] "Wenn es erlaubt ist einen Versuch zu machen den Begriff dieser Bewegung noch bestimmter zu fassen, so scheint im Allgemeinen eine der Ansicht von *Ampère* sich anschliessende Hypothese dem jetzigen Zustand der Wissenschaft am besten zu entsprechen. Denken wir uns die Körper aus Atomen gebildet [...]" (ibid. p. 35).

[157] "auch ist für unseren Zweck die Einsicht der Möglichkeit hinreichend, dass die Wärmeerscheinungen als Bewegungen gefasst werden können" (ibid. p. 36).

[158] "Das, was bisher Quantität der Wärme genannt worden ist, würde hiernach der Ausdruck sein erstens für die Quantität der lebendigen Kraft der Wärmebewegung, und zweitens für die Quantität derjenigen Spannkräfte in den Atomen, welche bei einer Veränderung ihrer Anordnung eine solche Bewegung hervorbringen können; der erstere Theil würde dem entsprechen, was bisher freie, der zweite dem, was latente Wärme genannt ist" (ibid. p. 35).

[159] "Die Erhaltung der Kraft würde bei dieser Bewegung so weit stattfinden, als bisher die Erhaltung der Quantität des Wärmestoffes erkannt ist, nämlich bei allen Erscheinungen der Leitung und Strahlung aus einem Körper zu dem andern, bei der Bindung und Entbindung von Wärme durch Aenderung des Aggregatzustandes" (ibid. p. 36).

[160] "Es bleibt die Wärmeentwickelung durch chemische Processe. Man hat dieselbe bisher für ein Freiwerden von Wärmestoff erklärt [...] Nach unserer Vorstellungsweise würde die bei chemischen Processen entstehende Wärme die Quantität der lebendigen Kraft sein, welche durch die bestimmte Quantität der chemischen Anziehungskräfte hervorgebracht werden kann, und das obige Gesetz würde der Ausdruck für das Princip von der Erhaltung der Kraft in diesem Falle werden" (ibid. pp. 36–7).

which compressed air is released and causes water to move, is adduced.[161] Helmholtz still stresses that Clapeyron's predictions, which were conform to measurements, are in agreement with those of Holtzmann, who establishes a relationship between heat and mechanical effect (Appendix F).[162]

Force equivalent of electricity

In the section 'force equivalent of electricity', we learn what is to be understood by tension force and living force in electrical phenomena. In the case of two charges, the intensity of the central force between them is proportional to the charges and inversely proportional to the square of the distance. The variation of the living force, when they are brought from a certain distance to another, is given by the tension force.[163]

Let us move on to one of the examples with electric current, the phenomenon discovered by Peltier. If electric current flows through two metals that are welded at the ends, it turns out that at the weld locations, there is heating or cooling. How can this phenomenon be looked at according to the law of conservation of force? If this law holds, there must be tension force and living force. Helmholtz considers that the heat from the battery corresponds to the tension force, and the heat in the circuit to the living force. The heat coming from the battery (chemical action) is given as $AJdt$, where A stands for electromotive force, J for current intensity and dt for a short time interval. (We will not penetrate into the meaning of this expression. We are interested in how Helmholtz constructs an equation with these terms and what he does with it.) The heat appearing in the circuit due to the current is given by Joule's or Lenz's law, i.e., WJ^2dt, where W represents resistance. The quantities of heat at the ends (where the metals are welded) are given as q_1dt and $-q_2dt$. The law of conservation of force leads to an equation, in which the first member concerns the force of tension and the second, the living force. In the first, the heat coming from the battery (force of tension) appears and in the other, the heat of the remaining circuit (living force),

$$\Delta \textit{force of tension} = \Delta \textit{living force}.$$

Thus, we have

161 "Ob bei der Erzeugung mechanischer Kraft Wärme verschwinde, was ein nothwendiges Postulat der Erhaltung der Kraft sein würde, ist noch niemals gefragt worden. Ich kann dafür nur einen Versuch von *Joule* anführen, der ziemlich zuverlässig zu sein scheint. Derselbe fand nämlich, dass die Luft bei dem Ausströmen aus einem Behälter [...] das umgebende Wasser um 4°.085 F. erkältete, sobald sie in die Atmosphäre ausströmte, also deren Widerstand zu überwinden hatte" (ibid. p. 37).

162 "übrigens folgt seine [Clapeyron] speciellere Formel für Gase, welche allein durch Vergleichung mit der Erfahrung unterstützt ist, auch aus der Formel von *Holtzmann*" (ibid. p. 39).

163 "Sind e, und e,, zwei electrische Massenelemente, deren Einheit diejenige ist [...] so ist die Intensität ihrer Centralkraft: $\varphi = -\frac{e_, e_{,,}}{r^2}$. Der Gewinn an lebendiger Kraft, indem sie aus der Entfernung R in die r übergehen, ist:

$-\int_R^r \varphi dr = \frac{e_, e_{,,}}{R} - \frac{e_, e_{,,}}{r}$ (ibid. p. 41).

$$AJdt = J^2Wdt + q_1dt - q_2dt \tag{2.9}$$

Dividing by *dt*, it follows[164]

$$AJ = J^2W + q_1 - q_2 \tag{2.10}$$

From this equation Helmholtz draws a consequence. Whether this consequence holds or not, he did not know. He claimed, more experimental research was needed.[165]

In sum, Helmholtz teachs us how to distinguish between force of tension and living force regarding the Peltier effect. The source of the circuit is taken as the former and all the rest as the latter. The law of force conservation provides the equality between them. At the end, this approach has a heuristic function. Whether or not the consequences of the law hold true would have to be tested.

Force equivalent of magnetism and electromagnetism

The first topic in this field is analogous to the first of the electricity section. Instead of two charges, we have now two magnets. The central force between two magnetic poles is proportional to their intensity and inversely proportional to the square of the distance, symbolically,

$$\varphi = -\frac{m_1m_2}{r^2} \tag{2.11}$$

The variation of the living force is linked with the force of tension in the same way *mutatis mutandis* as we have seen for electric charges.[166]

Let us now consider the motion of a magnet under the influence of electric current. In this case, we have a battery, which is at the source of the current, and the other part of the circuit, where the current flows, and a magnet, which moves. Helmholtz tells us what concerns force of tension and living force and then applies the conservation of force,

$$\Delta \textit{force of tension} = \Delta \textit{living force}.$$

---

164 "Denken wir uns einen hydroelectrischen constanten Strom, in dessen Leitungsdraht ein Stück eines anderen Metalls eingelöthet ist, dessen Löthstellen die Temperaturen *t'* und *t"* haben, so wird der electrische Strom während des Zeittheilchen *dt* in der ganzen Leitung die Wärme J^2 *W dt* erzeugen, ausserdem in der einen Löthstelle *q,dt* entwickeln, in der anderen *q,,dt* verschlucken. Ist A die electromotorische Kraft der hydroelectrischen Kette, also A J dt die chemisch zu erzeugende Wärme, so folgt aus dem Gesetze von der Erhaltung der Kraft [...]" (ibid. p. 57).

165 "Für beide Folgerungen sind mir bis jetzt noch keine messenden Versuche bekannt" (ibid. p. 58).

166 "Sind m, und m,, zwei magnetische Massenelemente, deren Einheit diejenige ist [...] so ist die Intensität ihrer Centralkraft: $\varphi = -\frac{m_,m_{,,}}{r^2}$. Der Gewinn an lebendiger Kraft beim Uebergang aus unendlicher Entfernung in die *r* ist.

$-\frac{m_,m_{,,}}{r}$ (ibid. p. 59).

He puts what concerns the battery on one side of the equation and the rest, on the other. The heat that develops in the battery and in the circuit during a short time interval dt were known, namely $AJdt$ and J^2Wdt (as seen above). Now Helmholtz expresses these quantities in mechanical terms, by multiplying them by the mechanical equivalent of heat. Thus, symbolizing this by a, we have $aAJdt$, for the battery (force of tension), and aJ^2Wdt, for the flow of the electric current (living force). The term relating to the magnet (living force) is given by $J\frac{dV}{dt}dt$, where V is the potential with respect to the unit current. Putting the terms together we have[167]

$$aAJdt = aJ^2Wdt + J\frac{dV}{dt}dt \tag{2.12}$$

From this equation Helmholtz takes the value of J. Instead of $J = \frac{A}{W}$, as was then known, he obtains

$$J = \frac{A - \frac{1}{a}\frac{dV}{dt}}{W} \tag{2.13}$$

The new term $\frac{1}{a}\frac{dV}{dt}$ is interpreted as a new driving force, the force of the induction current.[168]

In sum, the sense of force equivalent in electromagnetism is analogous to what we saw for electricity. The 'force equivalent' is what in this kind of phenomena corresponds to the tension force and to the living force. The law of conservation of force gives us the equation between the two forces. This equation has then a heuristic function.

On living beings

The final and briefest part of the book concerns living beings. Plants have a large amount of chemical tension forces and absorb only one form of living forces during growth, the radiation from sunlight. There were no indications, however, to express the equivalent of force in this field.[169] Regarding animals, there were more indications. They expend a certain amount of chemical tension forces and produce motion

[167] "Bewegt sich ein Magnet unter dem Einfluss eines Stromes, so muss die lebendige Kraft, die er dabei gewinnt, geliefert werden aus den Spannkräften, welche in dem Strome verbraucht werden. Diese sind während des Zeittheilchens dt nach der schon oben gebrauchten Bezeichnungsweise $A\ J\ dt$ in Wärmeeinheiten, oder $a\ A\ J\ dt$ in mechanischen, wenn a das mechanische Aequivalent der Wärmeeinheit ist. Die in der Strombahn erzeugte lebendige Kraft ist $a\ J^2\ W\ dt$, die vom Magneten gewonnene $J\ dV/dt$, wo V sein Potential gegen den von der Stromeinheit durchlaufenen Leiter ist. Also [...]" (ibid. pp. 61–2).

[168] "Wir können die Grösse $(1/a)\ (dV/dt)$ als eine neue electromotorische Kraft bezeichnen, als die des Inductionsstromes " (ibid. p. 62).

[169] "Vornehmlich wird in ihnen eine mächtige Quantität chemischer Spannkräfte deponirt, deren Aequivalent uns als Wärme bei der Verbrennung der Pflanzensubstanzen geliefert wird. Die einzige lebendige Kraft, welche dafür nach unseren bisherigen Kenntnissen [...] sind die chemischen Strahlen des Sonnenlichtes. Es fehlen uns indessen noch alle Angaben zur näheren Vergleichung der Kraftäquivalent, welche hierbei verloren gehen und gewonnen werden" (ibid. p. 66).

and heat, but still the indications were not precise enough for the discussion of the conservation of force, according to Helmholtz.[170]

Conclusion

In dealing with phenomena, Helmholtz has a conceptual framework that he applies to the observable and non-observable phenomena. By observable I mean the falling motion, the two electric charges or the two magnets poles considered above. In cases such as electric currents or heat phenomena, tension forces and living forces are not observable. In such cases, Helmholtz uses observables and interprets them, e.g. the battery that powers the circuit is subsumed by tension force and what follows by living force; latent heat corresponds to the tension force and sensible heat to the living force. The justification for using this conceptual framework is provided by the theoretical construction presented at the beginning of the book, which comes next.

Foundation of the approach

The core of Helmholtz's approach lies in the thesis, there are two ultimate forces in nature whose sum is constant. The argumentation behind the thesis starts as follows. The task of the theoretical part of science consists of the search for the ultimate causes of phenomena, according to the law of causality.[171] This means that if the causes by which phenomena are explained are changeable, one must continue the search until the ultimate causes are found.[172]

What are the ultimate causes?

To characterize ultimate causes, Helmholtz starts from the relation of the subject of scientific knowledge to external objects, that is, how objects in science are regarded from this primordial point of view. Science, Helmholtz continues, considers the objects of the external world according to a double abstraction, matter and force.[173] Matter as such has no effect on our sense organs. The effects that objects have on

[170] "Für die Thiere haben wir schon einige, nähere Anhaltpunkte. Dieselben nehmen die complicirten oxydablen Verbindungen, welche von den Pflanzen erzeugt werden, und Sauerstoff in sich auf, geben dieselben theils verbrannt als Kohlensäure und Wasser, theils auf einfachere Verbindungen reducirt wieder von sich, verbrauchen also eine gewisse Quantität chemischer Spannkräfte und erzeugen dafür Wärme und mechanische Kräfte [...] so reducirt sich die Frage nach der Erhaltung der Kraft ungefähr auf die [...] Diese Frage kann nach den Versuchen von *Dulong* und *Despretz* wenigstens annähernd bejaht werden" (ibid. p. 66).

[171] "Der theoretische Theil derselben sucht dagegen, die unbekannten Ursachen der Vorgänge aus ihren sichtbaren Wirkungen zu finden; er sucht dieselben zu begreifen nach dem Gesetze der Causalität" (ibid. p. 13).

[172] "Das endliche Ziel der theoretischen Naturwissenschaften ist also, die letzten unveränderlichen Ursachen der Vorgänge in der Natur aufzufinden" (ibid. p. 13).

[173] "Die Wissenschaft betrachtet die Gegenstände der Aussenwelt nach zweierlei Abstractionen: einmal ihrem blossen Dasein nach, abgesehen von ihren Wirkungen auf andere Gegenstände oder unsere Sinnesorgane; als solche bezeichnet sie dieselben als *Materie* [...] Wenn wir also den Begriff der Materie in der Wirklichkeit anwenden wollen, so dürfen wir dies nur, indem wir durch eine zweite Abstraction demselben wiederum hinzufügen, wovon wir vorher abstrahiren wollten, nämlich das Vermögen Wirkungen auszuüben, d. h. indem wir derselben Kräfte zuertheilen" (ibid. p. 14).

us are forces.[174] It is therefore erroneous, Helmholtz continues, to consider matter real and force not.[175] Matter and force are abstractions.[176] Furthermore, they are inseparable,[177] because it is by means of force that we infer the existence of matter.[178]

From this follows a first specification of the task of theoretical science. Since the existence of matter is inferred from force, the search for ultimate causes is the search for immutable forces.[179] Now, by the sense of force, immutable forces is what reaches our sense organs in an invariable way in time. But by the same sense of force, what reaches us in a constant way reflects what is permanent in matter. Thus, Helmholtz takes 'immutable forces' to mean 'indestructible qualities of matter'. These qualities had been determined by chemistry. The indestructible qualities of matter are the elements.[180]

If we think of the world, says Helmholtz, as consisting of elements, which by themselves do not change, any change in the world has to be spatial. It is therefore movement.[181] Since any change can only come from motion, phenomena must be traced back to movements of matter. The driving force of these motions can only depend on spatial relations.[182] To characterize these driving forces, Helmholtz imagines a bounded space where there are at least two bodies.[183] The change in the spatial

[174] "Das Dasein der Materie an sich ist uns also ein ruhiges, wirkungsloses [...] Qualitative Unterschiede dürfen wir der Materie an sich nicht zuschreiben, denn wenn wir von verschiedenartigen Materien sprechen, so setzen wir ihre Verschiedenheit immer nur in die Verschiedenheit ihrer Wirkungen d. h. in ihre Kräfte" (ibid. p. 14).

[175] "Ebenso fehlerhaft ist es, die Materie für etwas Wirkliches, die Kraft für einen blossen Begriff erklären zu wollen, dem nichts Wirkliches entspräche" (ibid. p. 14).

[176] "beides sind vielmehr Abstractionen von dem Wirklichen, in ganz gleicher Art gebildet; wir können ja die Materie eben nur durch ihre Kräfte, nie an sich selbst, wahrnehmen" (ibid. p. 14).

[177] "Es ist einleuchtend, dass die Begriffe von Materie und Kraft in der Anwendung auf die Natur nie getrennt werden dürfen" (ibid. p. 14).

[178] "Die Gegenstände der Natur sind aber nicht wirkungslos, ja wir kommen überhaupt zu ihrer Kenntniss nur durch die Wirkungen, welche von ihnen aus auf unsere Sinnesorgane erfolgen, indem wir aus diesen Wirkungen auf ein Wirkendes schliessen" (ibid. p. 14).

[179] "Wir haben oben gesehen, dass die Naturerscheinungen auf unveränderliche letzte Ursachen zurückgeführt werden sollen; diese Forderung gestaltet sich nun so, dass als letzte Ursachen der Zeit nach unveränderliche Kräfte gefunden werden sollen" (ibid. p. 14).

[180] "Materien mit unveränderlichen Kräften (unvertilgbaren Qualitäten) haben wir in der Wissenschaft (chemische) Elemente genannt" (ibid. p. 15).

[181] "Denken wir uns aber das Weltall zerlegt in Elemente mit unveränderlichen Qualitäten, so sind die einzigen noch möglichen Aenderungen in einem solchen System räumliche, d. h. Bewegungen" (ibid. p. 15).

[182] "die äusseren Verhältnisse, durch welche die Wirkung der Kräfte modificirt wird, können nur noch räumliche sein, also die Kräfte nur Bewegungskräfte, abhängig in ihrer Wirkung nur von den räumlichen Verhältnissen. Also näher bestimmt: Die Naturerscheinungen sollen zurückgeführt werden auf Bewegungen von Materien mit unveränderlichen Bewegungskräften, welche nur von den räumlichen Verhältnissen abhängig sind" (ibid. p. 15).

[183] "Räumliche Verhältnisse sind nur möglich gegen abgegrenzte Raumgrössen, nicht gegen den unterschiedslosen leeren Raum. Bewegung kann deshalb in der Erfahrung nur vorkommen als Aenderung der räumlichen Verhältnisse wenigstens zweier materieller Körper gegen einander" (ibid. p. 15).

relationship of these bodies can only be approximation or distancing, according to Helmholtz[184] If the change can only be approximation and distancing, it is understandable that Helmholtz comes to the conclusion that the forces that cause motion are either attractive or repulsive.[185]

Thus, Helmholtz arrives at the complete characterization of the task of theoretical science: explanation of phenomena by means of attractive or repulsive forces, which are directed along the line connecting the masses and dependent on distance between these.[186]

The task of physics is then to explain phenomena by these forces, called ultimate forces. In the case of falling bodies, the ultimate forces are found by observation in the sense that the body at a certain height (tension force) and the falling motion (living force) are observable. In the case of heat this is no longer the case. If one, however, admits that ultimate forces exist (which is justified conceptually) and accepts the task of physics (to explain phenomena by ultimate forces), it remains to imagine what is going on in the phenomena. This imagining act plays, however, a secondary role. It does not matter what motion the heat consists of but rather that heat can be thought of as motion. The same holds *mutatis mutandis* for the other cases. Indeed, we are taught about the correspondence between the ultimate forces and observable components of phenomena (battery, magnet, etc.). Hence, we have what we need to apply the law of conservation of force.

The advantages of Helmholtz's approach, following the author, is in the greater explanatory capacity of thermal phenomena in comparison with the theory of substantial heat and in the heuristic character of the equations he arrived at concerning electricity and electromagnetism.

2.4.2 Conclusion

Helmholtz assumes that no lasting force can exist out of nothing. Therefore, the force has to come from something. What forces exist and where they come from appears in the philosophical foundation of the theory. The forces that exist are taken from a thought experiment: two (chemical) elements isolated from the rest. This

184 "Bewegungskraft, als ihre Ursache, also auch immer nur erschlossen werden für das Verhältniss mindestens zweier Körper gegen einander, sie ist also zu definiren als das Bestreben zweier Massen, ihre gegenseitige Lage zu wechseln" (ibid. p. 15).

185 "Eine Bewegungskraft, welche sie gegen einander ausüben, kann deshalb auch nur Ursache zur Aenderung ihrer Entfernung sein, d. h. eine anziehende oder abstossende [...] Die Kräfte, welche zwei Massen auf einander ausüben, müssen nothwendig ihrer Grösse und Richtung nach bestimmt sein, sobald die Lage der Massen vollständig gegeben ist. Durch zwei Punkte ist aber nur eine einzige Richtung vollständig gegeben, nämlich die ihrer Verbindungslinie; folglich müssen die Kräfte, welche sie gegen einander ausüben, nach dieser Linie gerichtet sein, und ihre Intensität kann nur von der Entfernung abhängen" (ibid. pp. 15–6).

186 "Es bestimmt sich also endlich die Aufgabe der physikalischen Naturwissenschaften dahin, die Naturerscheinungen zurückzuführen auf unveränderliche, anziehende und abstossende Kräfte, deren Intensität von der Entfernung abhängt" (ibid. p. 16).

experience does not allow internal change of the bodies because they are elements, nor does it allow rotation or translation of the elements outside the direction that connects them. What can be observed in such a case are movements of the elements toward or away from each other. From these movements Helmholtz infers that they are caused. (This is justified: these movements represent a force (*vis viva*) and by the adopted principle a lasting force must have a cause.) As there are *only* two elements, these must have the ability to attract or reject. If the elements have this ability, then the thought experiment can only have the result that it has. Therefore, it is not the thought experiment that leads us to believe that there can only be two ultimate forces in nature, but that these are presupposed in advance.

Admitted the ultimate forces, Helmholtz moves on to the application to phenomena. He tells us, the battery in a circuit represents or has tension force. We cannot observe the parts of the battery between which the tension would exist, and we have no other indications from the author about this. This assigning of 'tension force' to the battery and 'living force' to the rest of the circuit has a role, however. It allows us to write equations of the form

$$\Delta tension\,force = \Delta living\,force.$$

These equations then play a heuristic role. From them, Helmholtz derives new information about the phenomena. In one case it was confirmed, in the other, not. (Bevilacqua, 1993, pp. 331–2).

2.4.2.1 Mayer and Helmholtz

Mayer assumes that cause equals effect. It is up to the observer to determine, whether a causal relationship exists in a given phenomenon. If there is, then cause and effect are forces, whereby

$$force\,(cause) = force\,(effect)$$

Helmholtz arrives at an equation of this form through the ultimate forces,

$$\Delta tension\,force = \Delta living\,force.$$

What each of them takes from the equations is different. From the equation in terms of heat and motion, Mayer takes the mechanical equivalent of heat. Helmholtz does not calculate the equivalent. He uses the equations for heuristic purposes.

The mechanical equivalent of heat is the foundation of Mayer's conservation of force. Transformation of force is also referred to the equivalent: it is the expression of a relation of quantities. Helmholtz has no experimental foundation of his idea of conservation of ultimate forces in the cases where these are non-observable.

Helmholtz needs the conjecture 'heat is motion' to generalize to heat forces of mechanical character. Conjectures also appear in electricity and electromagnetism.

Mayer's forces are observable. They are heat that can be measured, electrical effects that can be detected, the joining of chemicals in given quantities, etc. Hence, Mayer has several forms of force, whereas Helmholtz has the two ultimate forces.

2.4.2.2 Joule and Helmholtz

Helmholtz used Joule's experiments to show that heat is motion. On this point they both agree, heat is motion. Joule even suggests the hypothesis of a kind of motion of which heat would consist. The proper motion of heat is of no interest to Helmholtz. For him, it is enough that heat can be thought of as motion. This is sufficient, although also necessary to generalize to heat the forces of motion and tension.

Joule performed experiments to show that heat was motion, but still and mainly to arrive at equations that allowed him to determine the mechanical equivalent of heat. Helmholtz is a user of the mechanical equivalent of heat. The equivalent serves him to argue against the thesis 'heat is a substance' and to arrive at equations that play a heuristic role.

References

Ampère, A. M. (1822). Expériences relatives à de nouveaux phénomènes électro-dynamiques. *Annales De Chimie Et De Physique, 20*, 60–74.

Berthollet, C. L. (1809). Notes sur divers objects. *Mémoires de Physique et de Chimie de la Société d'Arcueil*. Tome II, 441–448. Johnson

Bevilacqua, F. (1993). Helmholtz' Ueber die Erhaltung der Kraft. In D. Cahan (Ed.), *Hermann von Helmholtz and the foundations of the Nineteenth-century Science* (pp. 291–333). University of California Press.

Carnot, S. (1824) *Réflexions sur la puissance motrice du feu*. Bachelier. (Rep. Paris: Éditions J. Gabay, 1990)

Colding, L. (1856) Nogle Sætninger om Kræfterne. Supplement to *Oversigt over det Kgl. Danske Videnskabernes Selskabs Forhandlinger 8*, 1–20. (Translation: Colding, L. (1972) Theses Concerning Force. In P. Dahl (ed.) *Ludvig Colding and the Conservation of Energy Principle*. Johnson Reprint Corporation.)

Colding, L. A. (1972) *Ludvig Colding and the Conservation of Energy Principle*. P. Dahl (ed. and transl.) Johnson Reprint.

Colladon, J.-D., & Sturm, C. F. (1828). Ueber die Zusammendrückbarkeit der Flüssigkeiten. *Annalen Der Physik, 88*, 161–197.

Davy, H. (1799) An Essay on Heat, Light, and Combinations of Light. In Davy, H. (1839–1840) *Collected Works*. J. Davy (ed.) Smith, Elder and Co., Vol. 2, pp. 2–86.

Delon, M. (1988). *L'idée d'énergie au tournant des Lumières (1770–1820)*. Pr Univ.

Faraday, M. (1832). Experimental researches in electricity. *Philosophical Transactions of the Royal Society of London, 122*, 125–162.

Haldat. (1807). Recherches sur la chaleur produite par le frottement. *Journal De Physique, De Chimie Et D'histoire Naturelle, 65*, 213–222.

Helmholtz, H. (1882). *Wissenschaftliche Abhandlungen I*. Barth.

Joule, J. P. (1838). Description of an Electro-Magnetic Engine. *Annals of Electricity* II, 122. In Joule (1884, pp. 1–3).

Joule, J. P. (1843). On the Calorific Effects of Magneto-Electricity, and on the Mechanical Value of Heat. *The London, Edinburgh, and Dublin Philosophical Magazine and Journal of Science* Series 3. Vol. XXIII, 263–276; 347–355; 435–443. In Joule (1884, pp. 123–159).

Joule, J. P. (1845a). On the Existence of an Equivalent Relation between Heat and the ordinary Forms of Mechanical Power. *Philosophical Magazine*, Series 3. Vol. XXVII, 205–207. In Joule (1884, pp. 202–205).

Joule, J. P. (1845b). On the Changes of Temperature produced by the Rarefaction and Condensation of Air. *Philosophical Magazine*, Series 3. Vol. XXVI, 369–383. In Joule (1884, pp. 172–189).

Joule, J. P. (1847a). On the mechanical equivalent of heat, as determined by the heat evolved by the friction of fluids. *Philosophical Magazine*, Series 3, 31, 173–176. In Joule (1884, pp. 277–283).

Joule, J. (1847b). Expériences sur l'identité entre le calorique et la force mécanique: détermination de l'equivalent par la chaleur dégagée pendant la friction du mercure. *Comptes Rendus*, 23 août. In Joule (1884, pp. 283–286).

Joule, J. P. (1850). On the mechanical equivalent of heat. *Philosophical Transactions of the R. S. of London 140*, 61–82. In Joule (1884, pp. 298–328).

Joule, J. P. (1884). *The scientific papers of James Prescott Joule.* Vol. 1. The Physical Society. (Rep. London: Dawsons, 1963.)

Kragh, H. (2009). Conservation and controversy: Ludvig Colding and the imperishability of "forces". *RePoSS: Research Publications on Science Studies, 4*, 1–27.

Leibniz, G. W. (1686). Brevis Demonstratio erroris memorabilis Cartesii, et aliorum circa legem naturae, secundum quam volunt a Deo eandem semper quantitatem motus conservari, qua et in re mechanica abutuntur. *Acta Eruditorum*, 161–163. In Leibniz, G. W. (1971) *Mathematische Schriften*, Vol. VI, C. I. Gerhardt (ed.), Hildesheim: G. Olms Verlag.

Martins, R. A. (2022). Joule's 1840 manuscript on the production of heat by voltaic electricity. *Notes Rec., 76*, 117–153.

Martins, R. A., & Silva, A. P. B. (2021). Joule's experiments on the heat evolved by metallic conductors of electricity. *Foundations of Science, 26*, 625–701.

Mayer, J. R. (1842). Bemerkungen über die Kräfte der unbelebten Natur. *Annalen Der Chemie Und Pharmacie, 42*, 233–240.

Mayer, J. R. (1845). *Die organische Bewegung in ihrem Zusammenhange mit dem Stoffwechsel.* Heilbronn: Drechsler.

Mayer, J. R. (1848). *Beiträge zur Dynamik des Himmels.* Heilbronn: Landherr.

Mayer, J. R. (1851). *Bemerkungen über das mechanische Aequivalent der Wärme.* Heilbronn: Landherr.

Rumford, B. C. (1798). An inquiry concerning the Source of the Heat which is excited by Friction. *Philosophical Transactions of the Royal Society of London, 88*, 80–102.

Seebeck, T. (1822–23). Magnetische Polarisation der Metalle und Erze durch Temperatur-Differenz. *Abhandlungen der Königlichen Akademie der Wissenschaften zu Berlin*, pp. 265–373.

Smith, C. W. (1976). Faraday as referee of Joule's Royal Society paper "On the Mechanical Equivalent of Heat" *Isis, 67*(3), 444–449.

Thomson, W. (1849). An account of Carnot's theory of the motive power of heat; with numerical results deduced from Regnault's experiments of steam. *Transactions of the R. S. of Edinburgh, 16*, 541–574.

Young, T. (1807). *Thomas Young's Lectures on Natural Philosophy and the Mathematical Arts*, I. (Rep. Bristol: Thoemmes, 2002)

Chapter 3
A New Concept

This chapter begins with the question that arises among experts in the theory of heat in the late 1840s: how to make this theory compatible with the new findings. Clausius resolved this issue. Thomson adhered then to the thesis that heat is motion and created the concept "mechanical energy". This concept not only covers what was properly mechanical, but also the activity due to heat, which had become a kind of motion. This is how the term energy appeared in the heat theory. A year later, Thomson distinguishes between static and dynamic reserves of mechanical energy, which led to the distinction between potential and kinetic energy. It is, finally, shown how the different theses on energy that had appeared since the 1840s emerge as characteristics of energy in the 1870s. This accumulation of meanings of the term energy made it more difficult to understand the concept.

3.1 Carnot's Theory and Joule's Experiments

During the 1847 meeting of the British Association for the Advancement of Science in Oxford, William Thomson learned about Joule's experimental work. The following year he published a paper in which he took a position on Carnot's theory and Joule's experiments. He accepts Joule's interpretation in cases where work gives rise to heat; but not in the reverse case, heat converting into work. He writes: "the conversion of heat (or *caloric*) into mechanical effect is probably impossible*". To this he adds a footnote:

> "This opinion seems to be nearly universally held by those who have written on the subject. A contrary opinion however has been advocated by Mr. Joule of Manchester; some very remarkable discoveries which he has made with reference to the *generation* of heat [...] seeming to indicate an actual conversion of mechanical effect into caloric. No experiment however is adduced in which the converse operation is exhibited" (Thomson, 1848, p. 315).

R. Lopes Coelho, *What Is Energy?*, History of Physics,
https://doi.org/10.1007/978-3-031-51855-3_3

He explains. The production of mechanical effect through heat, as happened in steam engines, was not to be seen as conversion, but as "transmission" of heat.[1] He exposes the sense of transmission through an analogy with falling water. Just as in a windmill, falling water gives rise to a mechanical effect, the passage of heat from the hot source to the cold source through the machine gives rise to a mechanical effect. The analogy continues: just as the falling water can be raised to its original height by outside work, so heat can be taken from the cold body and returned to the hot body by outside mechanical action.[2] This analogy, which already appeared in Carnot's book, takes heat as a substance.

There is something else that speaks for this theory. The 1848 article presents a new result based on Carnot's theory of heat: an absolute thermometric scale. In opposition to thermometry based on the volume variation of fluids, Thomson proposes a scale independent of substance.[3] This is founded on Carnot's theory by the following. Since according to Carnot, to a given temperature difference corresponds an amount of work, regardless of the substance used, then a mechanical effect can be used to define the temperature difference.[4] This is what Thomson uses. To a temperature difference of 1 degree (between any two values) he makes correspond the same mechanical effect.[5]

When Thomson wrote this paper, he had not yet read Carnot's (1824) book, *Reflections on the Motive Power of Fire*. He knew of it indirectly, through an 1834

---

[1] "In actual engines for obtaining mechanical effect through the agency of heat, we must consequently look for the source of power, not in any absorption and conversion, but merely in a transmission of heat" (Thomson 1848, p. 315).

[2] "Carnot [...] demonstrates that it is by the *letting down* of heat from a hot body to a cold body, through the medium of an engine (a steam-engine, or an air-engine for instance), that mechanical effect is to be obtained; and conversely, he proves that the same amount of heat may, by expenditure of an equal amount of labouring force, be *raised* from the cold to the hot body (the engine being in this case *worked backwards*); just as mechanical effect may be obtained by the descent of water let down by a waterwheel, and by spending labouring force in turning the wheel backwards, or in working a pump, water may be elevated to a higher level" (ibid. p. 315).

[3] "Although we have thus a strict principle for constructing a *definite* system for the estimation of temperature, yet as reference is essentially made to a specific body as the standard thermometric substance, we cannot consider that we have arrived at an *absolute* scale, and we can only regard, in strictness, the scale actually adopted as *an arbitrary series of numbered points of reference sufficiently close for the requirements of practical thermometry*. In the present state of physical science, therefore, a question of extreme interest arises: *Is there any principle on which an absolute thermometric scale can be founded*?" (ibid. p. 314).

[4] "The relation between motive power and heat, as established by Carnot, is such that *quantities of heat*, and *intervals of temperature*, are involved as the sole elements in the expression for the amount of mechanical effect to be obtained through the agency of heat; and since we have, independently, a definite system for the measurement of quantities of heat, we are thus furnished with a measure for intervals according to which absolute differences of temperature may be estimated" (ibid. p. 315).

[5] "The characteristic property of the scale which I now propose is, that all degrees have the same value; that is, that a unit of heat descending from a body A at the temperature T° of this scale, to a body B at the temperature (T−1)°, would give out the same mechanical effect, whatever be the number T. This may justly be termed an absolute scale, since its characteristic is quite independent of the physical properties of any specific substance" (ibid. p. 316).

article by Clapeyron.[6] He did, however, obtain a copy of the book and wrote another article on the theory the following year: *An account of Carnot's theory.*

This article begins by stating the two questions to which a complete theory of heat must answer:

- how one can produce mechanical effect from heat;
- how one can estimate the amount of heat required for a given effect.[7]

Now, says Thomson, Carnot's theory answers both questions:

- the thermal agent by which mechanical effect is obtained consists of the transfer of heat from one body to another of lower temperature[8];
- the perfect thermal machine, the machine imagined by Carnot, allows us to estimate the heat for a given mechanical effect.[9] Now, Thomson points out, citing the *Reflections*, that the amount of heat in a cycle remains invariant.[10] With this is connected the idea of heat being a substance. Being a substance, it could not be generated by any physical process.[11] This is in contrast to Joule's experiments, Thomson continues, referring to the heat generated by the magneto-electric machine.[12] Nevertheless, he still considers Carnot's theory to be adequate and

---

6 "*Réflexions sur la Puissance Motrice du Feu* [...] Having never met with the original work, it is only through a paper by M. Clapeyron, on the same subject [...] 1834 [...] that the author has become acquainted with Carnot's theory" (ibid. p. 313).

7 "The questions to be resolved by a complete theory of the subject are the following: (1) What is the precise nature of the thermal agency by means of which *mechanical effect* is to be produced, without effects of any other kind?

(2) How may the amount of this thermal agency necessary for performing a given quantity of work be estimated?" (Thomson, 1849, pp. 541–2).

8 "*The thermal agency by which mechanical effect may be obtained, is the transference of heat from one body to another at a lower temperature*" (ibid. p. 544).

9 "A *perfect* thermo-dynamic engine of any kind, is a machine by means of which the greatest possible amount of mechanical effect can be obtained from a given thermal agency [...] it is of primary importance to discover the criterion of a perfect engine. This has been done by CARNOT [...] *A perfect thermo-dynamic engine is such that [...] if an equal amount be spent in working it backwards, an equal reverse thermal effect will be produced*" (ibid. p. 545).

10 "Now the ordinarily-received, and almost universally-acknowledged, principles with reference to "quantities of caloric" and "latent heat," lead us to conceive that, at the end of a cycle of operations, when a body is left in precisely its primitive physical condition, if it has absorbed any heat during one part of the operations, it must have given out again exactly the same amount during the remainder of the cycle. The truth of this principle is considered as axiomatic by CARNOT" (ibid. p. 542) (See Carnot, 1824, p. 37).

11 "all those assumptions depending on the idea that heat is a *substance*, invariable in quantity; not convertible into any other element, and incapable of being *generated* by any physical agency" (ibid. p. 543).

12 "The extremely important discoveries recently made by Mr JOULE of Manchester, that heat is evolved in every part of a closed electric conductor, moving in the neighbourhood of a magnet, and that heat is *generated* by the friction of fluids in motion, seem to overturn the opinion commonly held that heat cannot be *generated*, but only produced from a source, where it has previously existed either in a sensible or in a latent condition" (ibid. p. 543).

thus, its fundamental principle, the invariability of the quantity of heat.[13] He gives an account of Joule's claim, to abandon Carnot's theory, but notes that this would give rise to numerous difficulties.[14] He then concludes that more experimental research is needed.[15]

In short, Joule maintained that he had proved the conversion of heat into work and work into heat. Thomson only accepts the latter. By virtue of this, rejecting Carnot's theory would be a mistake because this is a complete theory of heat.

Carnot theory and Joule experiments become compatible

In 1850, Clausius makes Carnot's theory and Joule's experiments compatible. He agrees that there is no experimental proof of the conversion of heat into work. He argues, however, that the experiments of the 1840s had shown that one could produce heat through work. (With this Thomson agreed.) Now, if heat could be produced through work, then heat could not be a substance. By the alternative of the time—heat being a substance or motion—it remained that heat was motion. What kind of motion was heat then? This did not matter to Clausius. It was enough for him that heat could be thought of as motion. Admitting heat as motion does not result in the rejection of Carnot's theory. Clausius develops then a theory of heat based on Carnot and Joule.

The first fundamental proposition of this theory tells us that the mechanical equivalent of heat is valid:

> "whenever work is expended, a certain amount of heat appears and vice versa, a certain amount of heat is capable of giving rise to a given amount of work",[16] which originates from Joule.

The second states that "*the production of work corresponds as an equivalent to the passage of heat from a hot body to a cold body*",[17] which originates from Carnot.

---

[13] "the fundamental axiom adopted by CARNOT may be considered as still the most probable basis for an investigation of the motive power of heat" (ibid. p. 544).

[14] This is also Carnot's point of view: "les quantités de chaleur absorbées ou dégagées dans ses diverses transformations sont exactement compensées. Ce fait n'a jamais été révoqué en doute [...] Le nier, ce serait renverser toute la théorie de la chaleur, à laquelle il sert de base." (Carnot, 1824, p. 37).

[15] "It might appear, that the difficulty would be entirely avoided, by abandoning CARNOT's fundamental axiom; a view which is strongly urged by Mr JOULE [...] If we do so, however, we meet with innumerable other difficulties—insuperable without farther experimental investigation, and an entire reconstruction of the theory of heat, from its foundation. It is in reality to experiment that we must look—either for a verification of CARNOT's axiom, and an explanation of the difficulty we have been considering; or for an entirely new basis of the Theory of Heat" (ibid. p. 545).

[16] "*daß in allen Fällen, wo durch Wärme Arbeit entstehe, eine der erzeugten Arbeit proportionale Wärmemenge verbraucht werde, und daß umgekehrt durch Verbrauch einer ebenso großen Arbeit dieselbe Wärmemenge erzeugt werden könne*" (Clausius, 1850, p. 373).

[17] "*Carnot* hat, wie schon oben erwähnt wurde, angenommen, *daß der Erzeugung von Arbeit als Aequivalent ein bloßer Uebergang von Wärme aus einem warmen in einen kalten Körper entspreche, ohne daß die Quantität der Wärme dabei verringert werde.* Der letzte Theil dieser Annahme, nämlich daß die Quantität der Wärme unverringert bleibe, widerspricht unserem früheren Grundsatze, und

(These two propositions would become 'the laws of thermodynamics'.) In the course of the study, Clausius shows how to approach thermodynamic phenomena based on these principles.

He also shows the convergence of Joule and Carnot regarding the mechanical equivalent of heat. Clausius argues as follows. According to Joule's experiments, the mechanical equivalent of heat is 460, 438 or 425 kg m. The conformity of these values to each other justifies the first fundamental proposition (the equivalence between work and heat) (Clausius, 1850, pp. 523–4). According to Carnot's theory, the equivalent of work to the unit of heat is 421 kg m. The convergence of this value with the mechanical equivalent of heat proves the truth of Carnot's fundamental law in the form given by the second fundamental proposition.[18]

3.2 Thomson's Mechanical Energy

March 1851: some of Thomson's views had changed. He embraces the dynamic theory of heat: heat is motion. The article, *On the dynamical theory of heat; with numerical results deduced from Mr. Joule's equivalent of a thermal unit [...]* begins with a historical reference: Humphrey Davy had been the founder of the dynamic theory of heat. This is justified by Davy's experiment of melting two pieces of ice by friction and two passages of the same text: caloric does not exist, heat is motion.[19] This is followed by a reference to Mayer and Joule:

> "The recent discoveries made by MAYER and JOULE, of the generation of heat through the friction of fluids in motion, and by the magneto-electric excitation of galvanic currents, would, either of them be sufficient to demonstrate the immateriality of heat; and would so afford, if required, a perfect confirmation of SIR HUMPHREY DAVY's views" (Thomson, 1851a, p. 261).

Clausius' article, as well as Rankine's, are referred to by Thomson as recent contributions to the dynamic theory of heat: the authors would have reached considerable conclusions, through reasoning analogous to Carnot's, but based on a principle contrary to his fundamental axiom (ibid. p. 262).[20]

---

muß daher, wenn wir diesen festhalten wollen, verworfen werden. Der erste Theil dagegen kann seinem Hauptinhalte nach fortbestehen" (ibid. p. 500).

[18] "die Uebereinstimmung derselben mit der Zahl 421 bestätigt in gleicher Weise die Richtigkeit des Carnot'schen Grundsatzes, in der Form, welche er durch die Verbindung mit dem ersteren Grundsatze angenommen hat" (ibid. p. 524).

[19] "SIR HUMPHREY DAVY, by his experiment of melting two pieces of ice by rubbing them together, established [...] he concludes that heat consists of a motion excited among the particles of bodies [...] The Dynamical Theory of Heat, thus established by SIR HUMPHREY DAVY is extended [...]" (1851a, p. 261).

[20] Rankine imagined the following: *"each atom of matter consists of a nucleus or central point, enveloped by an elastic atmosphere, which is retained in its position by attractive forces, and that the elasticity due to heat arises from the centrifugal force of those atmospheres, revolving or oscillating about their nuclei or central points"* (Rankine, 1850, p. 276). This paper was published in 1851. I

Next, Thomson lays out the fundamental principles of the heat theory. These are two propositions, the first said to be Joule's and the second, Carnot and Clausius'.[21] The first one establishes the correspondence between heat and mechanical effect.[22] The second proposition tells us that the reversible machine is the one that produces the maximum possible work among all those working between the same temperatures.[23]

Conclusion

Thomson's point of view had changed compared to his 1849 paper. Admittedly, Rankine's and Clausius' articles had been published but none of them satisfy Thomson's requirement of 1849, the need for experimental evidence. This difficulty does not, however, cease to be present in 1851. In a footnote, he points out that the conversion of heat into work possessed only theoretical arguments; the experimental proof that doing work brings about a decrease in heat was lacking.[24]

Thomson claims that the discoveries made by Mayer and Joule would afford a confirmation of Davy's views, who is presented as the founder of the dynamic theory of heat. It is true that heat is not matter for Mayer, just as it is not for Davy. However, the reasons are different. For Mayer, heat is a force and a force is not matter because matter is characterized by being ponderable and impenetrable and force by the opposite (Appendix D). Therefore, heat being a force, it could not be material. Davy's point of view was different. At that time, heat was either a substance or motion. Now, heat could not be a substance, because it could be created by the friction of

will use this version as it is more complete. On the quantity of heat, he wrote: "According to this hypothesis, *quantity of heat* is the *vis viva* of the molecular revolutions or oscillations" (Rankine, 1851, p. 510); and on the temperature: "the condition of *equilibrium of heat* is, that the square of the velocity of vortical motion, divided by the coefficient of atmospheric elasticity, shall be the same for each atom. Of this quantity, therefore, and of constants common to all substances, *temperature* must be a function" (ibid. p. 521). The temperature function is also given explicitly (see ibid. p. 523).

[21] "The whole theory of the motive power of heat is founded on the two following propositions, due respectively to JOULE, and to CARNOT and CLAUSIUS" (Thomson, 1851a, p. 264).

[22] "Prop. I. (JOULE).—When equal quantities of mechanical effect are produced by any means whatever, from purely thermal sources, or lost in purely thermal effects, equal quantities of heat are put out of existence, or are generated" (Thomson, 1851a, p. 264).

[23] "Prop. II. (CARNOT and CLAUSIUS).—If an engine be such that, when it is worked backwards, the physical and mechanical agencies in every part of its motions are all reversed; it produces as much mechanical effect as can be produced by any thermo-dynamic engine, with the same temperatures of source and refrigerator, from a given quantity of heat" (ibid. p. 264).

[24] "in a letter which I received from Mr JOULE on the 8th of July 1847. "In PELTIER'S experiment on cold produced at the bismuth and antimony solder, we have an instance of the conversion of heat into the mechanical force of the current," which must have been meant as an answer [...] but it would require a proof that there is more heat put out of existence at the heated soldering than is created at the cold soldering, to make the "evidence" be *experimental*. That this is the case I think is certain, because the statements of § 16 [...] but it is still to be remarked, that neither in this nor in any other case of the production of mechanical effect from purely thermal agency, has the ceasing to exist of an equivalent quantity of heat been demonstrated otherwise than theoretically. It would be a very great step in the experimental illustration (or *verification*, for those who consider such to be necessary) of the dynamical theory of heat, to actually shew, in any one case, a loss of heat" (Thomson, 1851a, pp. 267–8).

blocks of ice. Therefore, it would have to be motion. Moreover, the statement 'heat is motion' is not valid for Mayer, because this would mean that we know the essence of heat, which, according to him, is unknown. Regarding to Joule, the situation is different. Like Davy, Joule also argued that heat was motion, because it could not be a substance; and it could not be a substance because its quantity varies.

The idea that Davy is the founder of the dynamic theory of heat due to the ice piece friction experiment is not justified. Indeed, Rumford had argued in 1798, based on the friction of metals, that heat could not but be motion. Furthermore, Davy changed his views shortly after the publication of the 1799 paper.[25]

A new concept emerged

In December 1851, Thomson publishes another paper on the dynamic theory of heat. Here a concept of energy appears. More exactly, he defines "mechanical energy of a body in a given state". Let us consider the steps that lead to the definition, which allows us to understand what this concept means.

A body that emits heat or changes its volume against resisting forces does work on the outside.[26] To the extent that a body performs a mechanical effect on the environment, it changes *its own work reserve* in the same proportion, says Thomson.[27] (With this statement, Thomson assumes conservation of magnitude.) If there is a body-work reserve, as Thomson says, it makes sense to ask what work a body has in reserve. The total work reserve of a body or, what is the same, the mechanical value of all the effects that a body can produce is called the total mechanical energy of the body.[28] What is then the total value of those effects? This was not determinable:

> "in our present state of ignorance regarding perfect cold, and the nature of molecular forces, we cannot determine this "total mechanical energy" for any portion of matter" (Thomson, 1851b, p. 475).

As the problem is due to the lack of knowledge about the state of absolute zero, Thomson defines mechanical energy by reference to a *standard state.* The mechanical energy of a body in a given state is then understood to be the mechanical value of the effects which the body would produce if it passed from that state to another state, taken for standard state or, inversely, the mechanical value required to bring the body

---

[25] "Less than 2 years after, namely in December, 1800, referring to these productions, he [Davy] designates them, "my infant chemical speculations"; and, considering them chiefly in the light of mere speculations [...]" (Davy, 1839–1840, II, p. 3).

[26] "A body which is either emitting heat, or altering its dimensions against resisting forces, is doing work upon matter external to it. The mechanical effect of this work, in one case, is the excitation of thermal motions, and in the other, the overcoming of resistances" (Thomson, 1851b, p. 475).

[27] "The body must itself be altering in its circumstances, so as to contain a less store of work within it, by an amount precisely equal to the aggregate value of the mechanical effects produced" (ibid. p. 475).

[28] "The total mechanical energy of a body might be defined as the mechanical value of all the effect it would produce, in heat emitted and in resistances overcome, if it were cooled to the utmost, and allowed to contract indefinitely or to expand indefinitely according as the forces between its particles are attractive or repulsive, when the thermal motions within it are all stopped" (ibid. p. 475).

from the standard state to the state in question.[29] (Thomson admits that the passages in one direction and the other are equivalent.)

In order to obtain a mathematical expression, Thomson refers the reader to the March communication. He said then, suppose a mass under a given pressure p, whose volume v expands to $v + dv$ and the temperature changes from t to $t + dt$, the amount of work done will be

$$p\,dv$$

and the amount of heat that must be given off will be given by

$$Mdv + Ndt.$$

Using the mechanical equivalent of the thermal unit, symbolized by J, Thomson expresses heat in mechanical units and determines the total external effect as[30]

$$-J\,(Mdv + Ndt) + p\,dv.$$

Now, if this expression is the measure of the total external effect produced by the body and the body's reserve of work varies with this effect in the same proportion, it follows that the variation of the body's work is the inverse of the effect produced by the body. Therefore, the change in mechanical energy of a body, symbolized by '*de*', is,[31]

$$de = J(Mdv + Ndt) - pdv \tag{3.1}$$

[29] "the "mechanical energy of a body in a given state," will denote the mechanical value of the effects the body would produce in passing from the state in which it is given, to the standard state, or the mechanical value of the whole agency that would be required to bring the body from the standard state to the state in which it is given" (ibid. p. 475).

[30] "Let us suppose a mass of any substance, occupying a volume v, under a pressure p uniform in all directions, and at a temperature t, to expand in volume to $v + dv$, and to rise in temperature to $t + dt$. The quantity of work which it will produce will be $p\ dv$; and the quantity of heat which must be added to it to make its temperature rise during the expansion to $t + dt$ may be denoted by M dv + N dt. The mechanical equivalent of this is J (M dv + N dt), if J denote the mechanical equivalent of a unit of heat. Hence the mechanical measure of the total external effect produced in the circumstances is (p − J M) dv − J N dt" (Thomson, 1851a, p. 269).

[31] "The volume and temperature being denoted respectively by v and t, let e be the mechanical energy, p the pressure [...] M dv + N dt may express the quantity of heat that must be added to the fluid mass, to elevate its temperature by dt, when its volume is augmented by dv. The mechanical value of the heat added to the fluid in any operation, or the quantity of heat added multiplied by J (the mechanical equivalent of the thermal unit), must be diminished by the work done by the fluid in expanding against resistance, to find the actual increase of mechanical energy which the body acquires. Hence [...] we have de = [...]" (Thomson, 1851b, p. 476).

Conclusion

Thomson introduces the concept of mechanical energy. This energy is attached to a body: it is its amount of work in reserve. This reserve is potential in the classical sense. It is something that the body can do, but not that it necessarily does. If it does, its reserve changes and changes in the same proportion as the work done outside the body. This means that the quantity of work in the system 'body and its neighborhood' is conserved during a change. Thus, the idea of conservation is presupposed in this definition.

The definition of mechanical energy also depends on the mechanical equivalent of heat. Without this value it would not have been possible to express heat in terms of work. The quantitative expression of the variation of the mechanical energy would not have been written as it was (previous equation).

3.2.1 Thomson's Stores of Mechanical Energy

In 1852, Thomson publishes a paper on mechanical energy, where he defends the thesis that there is a universal tendency to its dissipation. The argument is based on Carnot's machine, thermodynamics phenomena and Creative Power. According to Carnot's proposition (within the framework of the dynamic heat theory), there is a *loss* of mechanical energy if the transformation is not performed by a perfect machine.[32] As a matter of fact, perfect machines do not exist. Therefore, that loss will always take place. Now, Thomson adds that only Creative Power can create or annihilate mechanical energy. Thus, it follows then that the loss of mechanical energy available for man is not really a loss. It consists of a transformation, but this is dissipative.[33]

To explain the nature of this transformation, Thomson divides the *reserves* of mechanical energy into static and dynamic.[34] Examples of the first kind are: a quantity of weights at a certain height, an electrically charged body, a quantity of fuel; and of the second kind, masses in motion, a volume passed through by waves of light or radiant heat, thermal motion of particles in a body (not infinitely cold).[35] With

[32] "The object of the present communication is to call attention to the remarkable consequences which follow from Carnot's proposition, established as it is on a new foundation, in the dynamical theory of heat; that there is an absolute waste of mechanical energy available to man" (Thomson, 1852a, p. 139).

[33] "As it is most certain that Creative Power alone can either call into existence or annihilate mechanical energy, the "waste" referred to cannot be annihilation, but must be some transformation of energy" (ibid. p. 139).

[34] "To explain the nature of this transformation, it is convenient, in the first place, to divide *stores* of mechanical energy into two classes—*statical* and *dynamical*" (ibid. p. 139).

[35] "A quantity of weights at a height, ready to descend and do work when wanted, an electrified body, a quantity of fuel, contain stores of mechanical energy of the statical kind. Masses of matter in motion, a volume of space through which undulations of light or radiant heat are passing, a body

this distinction, a transformation would be non-dissipative, if the mechanical energy could be restored to its initial condition.

In transformations between work and heat a restoration only takes place with the perfect machine.[36] Thomson exposes the consequences of this. If heat, for example, is caused by friction, restoration is not possible.[37] In the diffusion of heat by conduction or in the absorption of radiant heat or light, restoration is equally impossible.[38] Moreover, Thomson also admits that restoration is probably impossible even for organized matter.[39] The consequence of this would be that our life on Earth would be limited in time.[40] [This idea had a great repercussion on the society of the second half of the nineteenth century (Neswald, 2006; Kragh, 2008).]

Conclusion

The thesis of universal dissipation of mechanical energy starts from three main elements: Creative Power, the Carnot machine and thermodynamic phenomena. Since the perfect thermodynamic machine does not exist, the processes are irreversible. This means that the value of the mechanical effect at the beginning and end of a process is different. In accordance with the definition of mechanical energy, it has to be said, that the mechanical energy available to man decreases. It is assumed that only Creative Power can annihilate or create mechanical energy. Then, there can be no diminution of energy. Thomson reconciles the two aspects through the dissipative transformation.

In this article, Thomson provides a justification for the conservation of energy, which he has presupposed in the previous paper: creation or annihilation of energy can only be accomplished by Creative Power, assuming, which Thomson has to, that Creative Power is not changing the amount of mechanical energy on its own. Thomson does not tell us, however, how he arrived at this idea. It is true that the idea is not new, since Joule had argued in 1843 that "the grand agents of nature are, by

having thermal motions among its particles (that is, not infinitely cold), contain stores of mechanical energy of the dynamical kind" (ibid. p. 139).

[36] "When heat is created by a reversible process, (so that the mechanical energy thus spent may be *restored* to its primitive condition) there is also a transference [...] of a quantity of heat" (ibid. pp. 139–140).

[37] "When heat is created by any unreversible process (such as friction,) there is a *dissipation* of mechanical energy [...]" (ibid. p. 140).

[38] "When heat is diffused by *conduction*, there is a *dissipation* of mechanical energy, and perfect *restoration* is impossible.

IV. When radiant heat or light is absorbed, otherwise than in vegetation, or in chemical action, there is a *dissipation* of mechanical energy, and perfect *restoration* is impossible" (ibid. p. 140).

[39] "Any *restoration* of mechanical energy, without more than an equivalent of dissipation, is impossible in inanimate material processes, and is probably never effected by means of organised matter, either endowed with vegetable life, or subjected to the will of an animated creature" (ibid. pp. 141–2).

[40] "Within a finite period of time past the earth must have been, and within a finite period of time to come the earth must again be, unfit for the habitation of man as at present constituted" (ibid. p. 142).

the Creator's fiat, *indestructible*" (Joule, 1843, p. 442) and again in 1845 and 1847.[41] Nevertheless, he also did not say where he knew this from. Therefore, Thomson's thesis on the conservation of mechanical energy has no scientific argument.

3.2.2 Concerning the Concept of Energy

The distinction between static and dynamic reserves of mechanical energy introduced in this paper matters because it will lead to the distinction between potential and kinetic energy. Thomson's distinction is based on observation, when he tells us that a grave at a certain height from the ground has mechanical energy of the static kind, and if the body is in motion, it has dynamic mechanical energy. When he however states that a not absolutely cold body contains a reserve of mechanical energy of dynamic type, the characterization has another foundation, heat being conceived as motion. It is this concept that has, as a consequence, that a body that is not absolutely cold has thermal motions between its particles. These movements are not observable. In sum, the division of mechanical energy into a static and dynamic type is based on observation and conjecture.[42]

3.3 Rankine's Actual and Potential Energy

In 1853, Rankine publishes the article *On the general law of the transformation of energy*. The term energy comes naturally from Thomson. From him also comes the distinction between mechanical energy of a static and dynamic type, which, however, Rankine will significantly alter. After presenting these concepts, Rankine draws the reader's attention to the point that the principles and reasoning of the article may seem "abstract and metaphysical".[43]

---

[41] Criticizing Clapeyron about the "loss of *vis viva*", Joule wrote: "Believing that the power to destroy [*vis viva*] belongs to the Creator alone, I entirely coincide with Roget and Faraday in the opinion, that any theory which, when carried out, demands the annihilation of force, is necessarily erroneous" (Joule, 1845b, p. 189).

In 1847, he wrote: "absolute destruction of living force cannot possibly take place, because it is manifestly absurd to suppose that the powers with which God has endowed matter can be destroyed any more than that they can be created by man's agency" (Joule, 1847, p. 269).

[42] In another article from the same year, in which Thomson used the distinction of mechanical effects into static and dynamic, it also appears that observation and conjecture are the basis of the characterization. "The elevation of temperature produced in a body by the incidence of radiant heat upon it is a mechanical effect of the dynamical kind, since the communication of heat to a body is merely the excitation or the augmentation of certain motions among its particles" (Thomson, 1852b, p. 109).

[43] "Abstract and metaphysical as the principles and reasoning of this paper may appear, they are of immediate practical utility" (Rankine, 1853, p. 110).

By energy is meant any affection of substance which constitutes a power capable of overcoming resistance or is measurable as such (Substance means a body or a set of bodies, as we will see later). The term energy now includes not only motion, mechanical power, chemical action, heat, light, electricity, magnetism, but also all known or unknown powers which are convertible into them or measurable by them.[44]

Thomson's concept 'mechanical energy' is no longer used by Rankine but only 'energy'. The distinction of mechanical energy into 'static and dynamic' is replaced by the distinction of energy into 'potential and actual'.[45] By *actual energy* is meant the affection of the substance, which is measurable, transferable or transformable and whose presence induces a change of state of the substance.[46] *Potential energy* is defined by the elements of measurement, namely, the force or effort giving rise to a change and the quantity changed.[47] Thus, for example, if a certain pressure is exerted on a certain volume of gas, the pressure is seen as the effort tending to make a change and the volume which tends to be changed. Thus, if the volume of the gas is changed by a small value dV, then the potential energy is given by the product $P.dV$, where P represents the pressure.[48]

If a change occurs, the actual energy disappears, being replaced by potential energy and, conversely, if the latter disappears, actual energy reappears in the same proportion.[49] This is the Rankine content of the conservation of energy: if one disappears, the other reappears in the same proportion. He *admitted the conservation as known*. It is now presented as the constancy of the sum of the Universe's energies, actual

---

[44] "In this investigation the term *energy* is used to comprehend every affection of substances which constitutes or is commensurable with a power of producing change in opposition to resistance, and includes ordinary motion and mechanical power, chemical action, heat, light, electricity, magnetism, and all other powers, known or unknown, which are convertible or commensurable with these" (ibid. p. 106).

[45] "All conceivable forms of energy may be distinguished into two kinds; actual or sensible, and potential or latent" (ibid. p. 106).

[46] "*Actual energy* is a measurable, transferable, and transformable affection of a substance, the presence of which causes the substance to tend to change its state in one or more respects" (ibid. p. 106). "by the occurrence of which changes, actual energy disappears, and is replaced by *Potential energy*" (ibid. p. 106).

[47] "*Potential energy*, which is measured by the amount of a change in the condition of a substance, and that of the tendency or force whereby that change is produced (or, what is the same thing, of the resistance overcome in producing it), taken jointly" (ibid. p. 106).

[48] "Let V denote one measurable state, condition, or mode of existence of the substance under consideration, whose magnitude increases when the kind of potential energy in question is developed.

Let U denote this potential energy.

Let P be the tendency or force whereby the state V tends to increase, which is opposed by an equal resistance.

Then when the state V undergoes a small increase dV, the potential energy developed or given out is $dU = PdV$" (ibid. p. 106).

[49] "If the change whereby potential energy has been developed be exactly reversed, then as the potential energy disappears, the actual energy which had previously disappeared is reproduced" (ibid. p. 106).

and potential.[50] The novelty of the paper is then the law of energy transformation, which concerns the actual and potential energy[51]:

> "we obtain, for the total energy, actual and potential, acquired by the substance, in consequence of the changes of total actual energy from Q to Q + dQ, and of state, from V to V + dV, the formula
>
> $$d\psi = \sum dQ - \sum dU$$
>
> This equation is the complete expression of the *law of the transformation of energy*, from any one actual form Q to any one potential form U, developed by increase of the state V" (Rankine, 1853, p. 108–9).

In a second step, Rankine includes various forms of actual and potential energy. He writes then:

> "This equation is the complete expression of the law of the mutual transformation of actual and potential energy of all possible kinds" (ibid. p. 109).

Here we have two forms of energy—actual and potential—and many kinds of energy—heat, electricity, electromagnetism, etc. All these kinds of energy consist of those two forms.

Conclusion

In 1853, Rankine takes conservation of energy for granted. His novelty concerns the transformation of energy. This transformation occurs between two forms of energy which have a quantitative reciprocal relationship: if one increases, the other decreases. For this reason, they also justify the conservation of energy.

The designations of the energies can be related to the Aristotelian theory of act and potency. According to this, 'actual' refers to what is acting, happening and potential, to what may happen. Rankine's use of the terms actual and potential conforms to this sense.

3.3.1 Thomson Adopts Rankine's Concept

In the next year, Thomson adopts Rankine's concepts. A body has mechanical energy, says Thomson, when it is in motion or able to go into motion without external assistance.[52] The former can be called dynamic or actual energy and the latter potential

---

[50] "The law of the conservation of energy is already known, viz. that the sum of the actual and potential energies in the universe is unchangeable" (ibid. p. 106).

[51] "The object of the present investigation is to find the law of the *transformation of energy*, according to which all transformations of energy between the actual and potential states take place" (ibid. p. 106).

[52] "Any piece of matter, or any group of bodies, however connected, which either is in motion, or can get into motion without external assistance, has what is called mechanical energy" (Thomson, 1854, p. 34).

energy.[53] The term 'potential' has been completely adopted by Thomson. The term 'actual' has dynamic as a variant. (It is the latter term that will lead to 'kinetic', giving rise to what we still use, potential and kinetic energy.) Let's move on to the interpretation of phenomena.

> "A stone at a height [...] has potential energy. If the stone be let fall, its potential energy is converted into actual energy during its descent, exists entirely as the actual energy of its own motion at the instant before it strikes, and is transformed into heat at the moment of coming to rest on the ground" (Thomson, 1854, p. 34).

In the 1852 paper, a stone at a certain height had mechanical energy of a static kind; when dropped it was left with a reserve of mechanical energy of a dynamic kind. The change from static to dynamic energy occurred with the beginning of the movement, because once in motion, there is no longer static energy. In 1854, this changed. If dropped, the potential energy of the stone converts to actual energy during the descent. This differs from the previous characterization since energy of static kind did not convert into dynamic during the descent. It ended when the stone went into motion. Hence, if Thomson had maintained energy of a *static kind*, he would not have been able to speak of conversion. 'Static' has no gradation. The stone is either at rest or not. Dynamic, on the contrary, can be used during the motion.

The mechanical equivalent of heat (Joule, 1850) is interpreted through the new terminology: the actual or dynamic energy of heat capable of raising the temperature of a pound of water by 1°F is an exact equivalent of the potential energy of a pound of matter at a height of 772 feet.[54]

Based on Joule's experiments, Thomson says, set off his "speculations."[55] Here are two examples of speculations mentioned in the paper: war and the principle of the Universe. The potential energy of war would be contained in the reserves of gunpowder, which artillery and infantry use, and in supplies for men and horses. The artillerymen, pedestrians and other participants are seen as means or tools, through which the potential energy is used for the intended purposes.[56] As in 1852, Thomson

---

[53] "The energy of motion may be called either "dynamical energy" or "actual energy". The energy of a material system at rest, in virtue of which it can get into motion, is called "potential energy."" (ibid. p. 34).

[54] "Mr Joule, by a series of well planned and executed experiments, ascertained that a pound of water would have its temperature increased by 1° (Fahrenheit) if it kept all the heat that would be generated by its descent through 772 feet; that is, the "actual" or "dynamical" energy of as much heat as raises the temperature of a pound of water 1° is an exact equivalent for the potential energy of a pound of matter 772 feet above the ground" (ibid. p. 35).

[55] "These researches, with the theory of animal heat and motion [...] due to the same penetrating investigator [Joule], have afforded to the author of the present communication the chief groundwork for his speculations" (ibid. p. 35).

[56] "The potential energy of war is contained in the stores of gunpowder and food brought into the field. The gunpowder carried by artillery and infantry contains all the potential energy ordinarily brought into action by those two arms of the service. The men's food, and the forage for the horses, contain the stores of potential energy drawn upon in a charge of cavalry. Artillerymen, foot soldiers, sailors, steamers with their engines, guns swords, are only means and appliances by which the potential energy contained in the stores of gunpowder and food is directed to strike the blows by which the desired effects are produced" (ibid. p. 35).

believes that there will be an end to life in the world.[57] He now puts forward a hypothesis of the beginning of the universe: in the beginning, there would have been only the potential energy of gravitation, which would therefore be the antecedent of motion, heat and light of the Universe.[58]

Conclusion

Thomson adopted the term potential energy, which replaces his static mechanical energy and opens up the possibility of a transformation of energy between potential and actual forms. He also adopted the term actual, but alongside the term dynamic.

Thomson interpreted Joule's paddle wheel experiment by means of the new terminology. He wrote:

> "the "actual" or "dynamical" energy of as much heat as raises the temperature of a pound of water 1° is an exact equivalent for the potential energy of a pound of matter 772 feet above the ground" (ibid. p. 35).

Let's go back to Joule's result to analyze how it fits into the new terminology. Joule wrote:

> "the quantity of heat capable of increasing the temperature of a pound of water (weighed in vacuo, and taken at between 55° and 60°) by 1°F requires for its evolution the expenditure of a mechanical force represented by the fall of 772 lb. through the space of one foot" (Joule, 1850, p. 82).

This was the result he obtained with his machine. If this is in motion, we can observe two falling bodies and the movement of the paddles that agitate the water. The falling bodies have a certain speed, which Joule determined.[59] Therefore, we can write down the actual/dynamic energy of these bodies. Therefore, we have the actual/dynamic energy of these bodies and that which corresponds to the heat that raises a pound of water by one degree Fahrenheit. If we accept these two actual/dynamic energies, then the potential energy of a pound of matter 772 feet above the ground is *not* an exact equivalent of the actual or dynamic energy that corresponds to the heat of 1°F, as Thomson said.

3.3.2 Concerning the Concept of Energy

The interpretation of phenomena in terms of actual and potential energy in the non-speculative part of the article—the stone, the flow of water, Joule's experiment—shows that the terms *transformation* and *conversion* are used synonymously. This undifferentiated use of the terms eliminates the distinction between the term

---

[57] "we find that the end of this world as a habitation for man, or for any living creature or plant at present existing in it, is *mechanically inevitable*" (ibid. p. 37).

[58] "*the potential energy of gravitation may be in reality the ultimate created antecedent of all the motion, heat, and light at present in the universe*" (ibid. p. 40).

[59] "Velocity of the weights in descending, 2.42 inches per second." (Joule 1850, p. 65).

connected to forms of energy, which comes from Mayer, and that connected to heat as motion, which comes from Joule. This favors the path to eclecticism, as we shall see.

3.4 The Science of Energetics

In 1855, Rankine lays the foundations of an energy-based science. The article, *Outlines of the science of energetics*, consists of two parts: the first is about theorizing in physics (what kinds of theories exist), and the second part is dedicated to science of energetics. I will begin with the latter. This science is presented through basic concepts and axioms. These concepts, however, differ from what was used in the domain. Some known propositions about energy appear in a new form as they are expressed through the new terminology. The intellectual effort that is required of anyone joining the energetics is justified by the philosophy of science presented in the first part of the article.

The object of energetics is material bodies and physical phenomena in general.[60] The basic concepts used are of Aristotelian origin: *substance* and *accidents*. Work, energy and energy forms are then defined by means of these concepts.

'Substance' is the term used to refer to a body, a part of a body or a set of bodies.[61] 'Accident' refers to any variable state of a substance.[62] Accidents can be:

- absolute or relative,
- homogeneous or heterogeneous,
- active or passive.

An accident is said to be *absolute* or *relative*, according to whether the variable state of the substance depends on the condition of each part or on a relationship between parts of a substance.[63] They are *homogeneous* or heterogeneous, depending on whether or not they can be reduced to a unit or quantity.[64] Accidents can be *active* or *passive* according to whether they tend to vary an accident or are the varied accident. We turn next to the use of these terms.

Work is understood as the variation of a passive accident through an active one. Thus, for example, if the work consists of force times displacement, force is the active

---

[60] "a science whose subjects are, material bodies and physical phenomena in general, and which it is proposed to call the SCIENCE OF ENERGETICS" (Rankine, 1855, p. 214).

[61] "The term "*substance*" will be applied to all bodies, parts of bodies, and systems of bodies" (ibid. p. 214).

[62] "The term "*accident*" will be applied to every variable state of substances" (ibid. p. 214).

[63] "consisting in a condition of each part of a substance, how small soever, (which may be called an *absolute accident*), or in a physical relation between parts of substances, (which may be called a *relative accident*)" (ibid. pp. 214–215).

[64] "Accidents may be said to be *homogeneous* when the quantities expressing them are capable of being put together, so that the result of the combination of the different accidents shall be expressed by one quantity" (ibid. p. 215).

accident and displacement is the passive. If work is given by pressure times volume, pressure is the active and volume the passive accident. In general, the magnitude work is given by the integral

$$W = \int_{x_0}^{x_1} X dx \tag{3.2}$$

where x denotes the passive accident;

X is the active accident or effort, which causes the passive accident to vary;

W stands for the work done in increasing x, from x_0 to x_1 (Rankine, 1855, pp. 216–7).

The terms substance and work are used to define energy. Energy means the state of a substance capable of performing work.[65] If the capacity to do work depends on the substance itself (it is an absolute accident), it is called *actual energy*; if this capacity of the substance to do work depends on other substances (it is a relative accident) it is called *potential energy*.[66] Once the terminology has been presented, Rankine moves on to the axioms.

Axioms

The science of energetics is based on three axioms. The first states:

> "All kinds of Work and Energy are Homogeneous." (ibid. p. 218).

Given the definition of homogeneous accidents—these can be reduced to a unit or quantity—the axiom tells us that work and energy are expressed by the same magnitude. The consequence Rankine draws is that "*any kind of energy may be made the means of performing any kind of work*" (ibid. p. 218).

The second axiom states:

> "The Total Energy of a Substance cannot be altered by the Mutual Actions of its Parts" (ibid. p. 219).

If the energy of a substance cannot be altered by its parts, then, without external action, its energy is conserved. Rankine concludes from the axiom that "*all work consists in the transfer and transformation of energy alone*" (ibid. p. 219).

The third axiom states:

> "The Effort to Perform Work of a Given Kind, caused by a Given Quantity of Actual Energy, is the Sum of the Efforts caused by the Parts of that Quantity." (ibid. p. 220).

---

65 "The term "*energy*" comprehends every state of a substance which constitutes a capacity for performing work" (ibid. p. 217).

66 "'*Actual energy*' comprehends those kinds of capacity for performing work which consist in particular states of each part of a substance [...] that is, in an *absolute accident*, such as heat, light, electric current, *vis viva*. [...] "*Potential energy*" comprehends those kinds of capacity for performing work which consist in relations between substances, or parts of substances; that is, in *relative accidents*" (ibid. p. 217).

One part of the axiom tells us that the effort to do work is the sum of the efforts. The other part concerns the origin of these efforts. Let us consider an example. One kind of work is pressure times the volume on which the pressure acts. Pressure is therefore an effort. Applying the axiom, we have that pressure is caused by a given quantity of actual energy. Thus, we have that actual energy is at the origin of the pressure that creates potential energy. It follows that actual energy gives potential. In this way, the axiom appears related with the law of transformation of energy. Indeed, according to Rankine, the axiom is equivalent to this law (1853).[67]

In the last section of the article, it is noted to the reader that the results of the sections do not come from speculation, but from the generalization of a method of reasoning that would have already been successful in some particular domains of physics.[68] This is explained in Rankine's philosophy of physics, outlined in the first part of the article, as follows.

According to Rankine, the advance of scientific knowledge takes place in two phases:

- observation of phenomena and their relationships;
- establishment of the principles from which laws can be deduced for a class of phenomena.[69]

A set of principles with their systematically deduced consequences is called a physical theory.[70] In the construction of a theory, Rankine continues, two methods can be distinguished, the abstractive and the hypothetical.[71] The abstractive is characterized by describing phenomena and assigning names or symbols based on perception. (This is the method adopted for the science of energetics.) Following the hypothetical method, the class of phenomena that is the object of the theory is defined according to

---

67 "A law equivalent to this axiom, under the name of the "GENERAL LAW OF THE TRANSFORMATION OF ENERGY,"" (ibid. pp. 220–1).

68 "It is to be observed, that the preceding articles are not the results of a new and hitherto untried speculation, but are the generalised expression of a method of reasoning which has already been applied with success to special branches of physics" (ibid. p. 227).

69 "An essential distinction exists between two stages in the process of advancing our knowledge of the laws of physical phenomena; the first stage consists in observing the relations of phenomena [...] The second stage consists in reducing the formal laws of an entire class of phenomena to the form of a science" (ibid. p. 209).

70 "in discovering the most simple system of principles, from which all the formal laws of the class of phenomena can be deduced as consequences.

Such a system of principles, with its consequences methodically deduced, constitutes the PHYSICAL THEORY of a class of phenomena" (ibid. p. 209).

71 "Two methods of framing a physical theory may be distinguished [...]. They may be termed, respectively, the ABSTRACTIVE and the HYPOTHETICAL methods" (ibid. p. 210).

a conjecture about its nature.[72] This conjecture, in turn, may be objective or subjective, according to whether it is something that may exist but is imperceptible to us, or is simply a means of expressing phenomena. Mechanics is given as an illustration of a physical theory of abstractive method.[73]

Indeed, mechanics is not only the example of a theory of abstractive method, but also the only complete physical theory, according to Rankine. Herein lies the reason for the adoption of mechanical hypotheses in other physical theories.[74] These hypothetical theories proved useful in some areas, because they allowed complex phenomena to be reduced to simple laws of mechanics. Nevertheless, they also had drawbacks, namely, Rankine continues, by diverting attention away from phenomena not consistent with the hypotheses adopted.[75] This criticism is illustrated by the little attention given to frictional heat in the heat-substance theory, the phenomenon was not in line with the hypothesis.[76] All this justifies the author's choice in the construction of science of energetics.[77] Instead of imagining movements and forces that are hidden from us, he considers properties in various sets of phenomena in order to arrive at general principles.[78]

Conclusion

Rankine tells us that he did not adopt the hypothetical method in his construction of the science of energetics. This means that he does not imagine movements and forces that are hidden from us. It follows that we are not allowed to interpret actual and potential energy as being hidden particle motions and forces. It turns out that actual and potential energy comes from mechanical energy of dynamic and static kinds,

---

[72] "According to the ABSTRACTIVE method, a class of objects or phenomena is defined by describing [...] and assigning a name or symbol to, that assemblage of properties which is common to all the objects or phenomena composing the class, as perceived by senses, without introducing anything hypothetical. According to the HYPOTHETICAL method, a class of objects or phenomena is defined, according to a conjectural conception of their nature" (ibid. p. 210).

[73] "The principles of the science of mechanics, the only example yet existing of a complete physical theory, are altogether formed from the data of experience by the abstractive method" (ibid. p. 210).

[74] "The fact that the theory of motions and motive forces is the only complete physical theory, has naturally led to the adoption of *mechanical hypotheses* in the theories of other branches of physics" (ibid. p. 211).

[75] "It is well known that certain hypothetical theories [...] have proved extremely useful [...] The neglect of the caution already referred to [...] and a tendency has, consequently, often evinced itself to explain away, or set aside, facts inconsistent with these hypotheses" (ibid. p. 212).

[76] "Thus, the fact of the production of heat by friction, the basis of the true theory of heat, was long neglected, because inconsistent with the hypothesis of caloric" (ibid. p. 212).

[77] "Besides the perfecting of mechanical hypotheses, another and an entirely distinct method presents itself for combining the physical sciences into one system; and that is, by an *extension of the* ABSTRACTIVE PROCESS in framing theories" (ibid. p. 213).

[78] "Instead of supposing the various classes of physical phenomena to be constituted, in an occult way, of modifications of motion and force, let us distinguish the properties which those classes possess in common with each other [...] So shall we arrive at a body of principles, applicable to physical phenomena in general" (ibid. p. 213).

which are mechanical concepts. Here the question of whether actual and potential energy are to be interpreted as conjecture or not arises.

In 1855, potential and actual energy are based on the concept of accident. Accident is just a variable state of a substance. So actual and potential energy are variable states of a substance. The further question of whether there are hidden particle motions and forces, does not arise because it is outside of what is defined by accident. Moreover, it would be a mechanical conjecture, which is characteristic of the hypothetical method and contrary to the abstractive method that Rankine claims to have followed. Furthermore, this interpretation is also in accordance with the characterization of the terminology of energy science: it is purely abstract.[79] Thus, the conjectural character of actual and potential energy does not hold for Rankine.

Helmholtz and Rankine

Helmholtz defended the conservation of ultimate forces, the tension force and the living force. The decrease in one implies the increase of the other in the same proportion. In some cases, as in free fall, the bodies are observable. In others, as in heat, Helmholtz conjectured that corpuscles behaved in the same way.

Rankine generalizes actual and potential energy to all domains. He is not concerned with the conservation of energy, which he simply admits as known. Rather, he presents the general law of energy transformation, which includes only those two forms, whatever the domain.

There is, therefore, a parallelism between the pairs 'live force and tension force' and 'actual and potential energy'. If one of the terms of the pair increases, the other decreases so that either pair conforms to conservation. There is, however, also a difference. Helmholtz admits that the ultimate forces exist. According to Rankine's philosophy, Helmholtz followed the hypothetical method in the construction of his theory. On the contrary, Rankine followed the abstractive. Actual and potential energy refers only variable states of a substance but not a mechanical conjecture.

3.5 Concerning Kinetic and Potential Energy

The adjective kinetic in 'kinetic energy' was introduced by Thomson and Tait in an article, entitled *Energy*, published in a magazine aimed at the general public, *Good Words*, 1862. The authors explain that they prefer 'kinetic' energy to 'actual', because 'kinetic' indicates the form in which energy reveals itself, which is motion.[80] Moreover, its magnitude is calculated as a function of the mass and velocity of the

[79] "The peculiar terms which will be used in treating of the Science of Energetics are purely abstract". (ibid. p. 214).

[80] "[...] It had KINETIC or (as it has sometimes been called) actual energy. We prefer the first term, which indicates motion as the form in which the energy is displayed" (Thomson and Tait, 1862, p. 602).

body.[81] More exactly, it is given by half of the product of the mass and square of the velocity of the body.[82]

In 1883, Thomson explained to us the reason for introducing the adjective 'kinetics'. He had intended to reformulate the terminology of Mechanics: to use the term 'mechanics' for the science of machines and 'dynamics' for the science of forces. The usual division of mechanics into statics and dynamics would necessarily have to be changed. The term dynamics was then replaced by kinetic. This had the consequence that the expression 'actual or dynamic' became 'actual or kinetic'.[83] Thomson and Tait reduced then to 'kinetics' for the reason explained earlier. Another consequence of the reform concerns the concept 'mechanical equivalent of heat', which was called 'dynamic equivalent of heat'. This change was not adopted by the scientific community. The expression 'mechanical energy' was not changed to 'dynamic energy'. It simply disappeared, Thomson and Tait speak only of 'energy'.

Once we get to the expression we use today, let us briefly look at the conceptual development of these two forms of energy. Thomson created the concept of mechanical energy and made a distinction between mechanical energy of the static kind and dynamic kind. Mechanical energy of a body in a given state refers to the work reserve of the body in that state. The work reserve was information for us humans because it told us what work could be counted on: it was mechanical energy available for man. The distinction between the static and the dynamic kind refers to bodies or particles which are at rest or in motion. In cases where movements were not observed, a conjecture was made (in heat, electricity, etc.): there is either rest or movement of the very small parts of bodies (A third hypothesis is inconceivable).

Rankine realized that this is not just a way of systematizing the reserves of work. He introduces the concepts of potential and actual energy, which account for Thomson's reserves of mechanical energy and the conservation of energy. Unlike Thomson, (1852a, b) Rankine (1853) establishes a quantitative relationship between actual and potential energy: if one increases the other decreases. With this increase and decrease in the same proportion, energy is conserved. His novelty is then the general law of energy transformation, as being the mutual transformation of actual and potential energies in all domains, heat, electricity, etc.

The change from actual to kinetic performed by Thomson and Tait is only the result of a terminological reform. It does not change what Rankine had proposed but only emphasizes that actual energy is motion (Table 3.1).

---

[81] "Kinetic energy depends on motion; and observation shows that its amount in each case is calculable from the mass which moves and the velocity with which it moves" (ibid. p. 602).

[82] "there is particular advantage in taking as the exact expression, *one-half of the product of the moving mass and the square of its velocity in feet per second*" (ibid. p. 602).

[83] In 1883, Thomson wrote: "A few years later, in advocating a restoration of the original and natural nomenclature,—"mechanics the science of machines,"—"dynamics the science of force," I suggested (instead of statics and dynamics the two divisions of mechanics according to the then usual nomenclature) that statics and kinetics should be adopted to designate the two divisions of dynamics. At the same time I gave, instead of "dynamical energy," or "actual energy," the name "kinetic energy" which is now in general use to designate the energy of motion" (Thomson, 1884, p. 34).

Table 3.1 The mechanical origin of kinetic and potential

Mechanical energy	Energy	Energy
Static kind	Potential	Potential
Dynamic kind	Actual	Kinetic
Thomson (1852a, b)	Rankine (1853)	Thomson & Tait (1862)

3.6 Energetic Eclecticism

In the 1870s, Maxwell published his *Theory of Heat* several times. This work brings together Rankine's definition of energy, Thomson's definition of mechanical energy and the energy conservation principle in Rankine's and Mayer's form. These authors, however, are not mentioned. If they had been, the reader would know that these authors have thought about the subject in different ways. Since the reader does not have this information, he is led to think that everything that is gathered from these authors about energy is characteristic of it. This is what is meant by energetic eclecticism.

Energy of a body, according to Maxwell, means the capacity of the body to do work.[84] (Here we have Rankine's definition of energy.) The work that the body can do depending on its current situation is called intrinsic energy of the body.[85] (This energy is Thomson's 'mechanical energy'.) Its absolute value, that is, the total energy of the body, is not determinable, because it would be impossible, says Maxwell, to get a Carnot machine working between the temperature of the body and absolute zero. (The total mechanical energy is not determinable, according to Thomson.) Thus, intrinsic energy is determined in applications with respect to a standard state.[86] (This is Thomson's standard state.) For this reason, the energy of a body can be negative, but Maxwell notes that this only means that the energy of the standard state is higher than that of the body; an actual negative energy would be impossible.[87]

The principle of conservation of energy is introduced in the form, the sum of the potential and kinetic energy of all the bodies in a system will always remain the same.[88] (This formulation of the principle comes from Rankine.) Maxwell notes,

---

[84] "the energy of a body may be defined as the capacity which it has of doing work" (Maxwell, 1873, p. 90).

[85] "The Intrinsic energy of a body is the work which it can do in virtue of its actual condition, without any supply of energy from without" (ibid. p. 183).

[86] "we cannot determine experimentally the whole energy of the body. It is sufficient, however, for all practical purposes to know how much the energy exceeds or falls short of the energy of the body in a certain definite condition—for instance, at a standard temperature and a standard pressure" (ibid. pp. 183–4).

[87] "If the body in its actual state has less energy than when it is in the standard state, the expression for the relative energy will be negative. This, however, does not imply that the energy of a body can ever be really negative, for this is impossible. It only shows that in the standard state it has more energy than in the actual state" (ibid. p. 184).

[88] "the sum of the potential and kinetic energy of all the bodies of the system will always remain the same. This principle is called the Principle of the Conservation of Energy" (ibid. p. 92).

however, that it cannot be said that all energy is kinetic or potential, although we are unable to conceive of other forms.[89] In favor of the principle are the experiments: the principle has been shown experimentally correct in cases where energy takes the form of heat, magnetization, electricity, etc.[90] The principle is not necessarily true, says Maxwell, but which one cannot but take into account. In its most general form, it states[91]:

> "The total energy of any body or system of bodies is a quantity which can neither be increased nor diminished by any mutual action of these bodies, though it may be transformed into any of the forms of which energy is susceptible." (Maxwell, 1873, p. 92–93).

In this formulation, energy is a *quantity*. It is understandable then that Maxwell says that it can neither be increased nor decreased, instead of 'can neither be created nor destroyed'. The latter refers to something that we cannot destroy, so it must exist, which is not proper of a quantity, which can only increase, decrease or remain constant. At the final part of the formulation we read, however, 'it may be transformed'. To be transformed is not proper to a quantity, but rather refers to something that can go through metamorphoses.

The transformation of energy is connected with the concept of heat. There is no reason to believe that heat is a substance, says Maxwell, because it can be generated.[92] However, there is reason to believe that it is a form of energy, because heat can be generated by work and work can be done by heat. Furthermore, he adds, there is a numerical relationship: for every unit of work that disappears, there is a given amount of heat.[93]

Conclusion

'Forms of energy' signals two meanings. On the one hand, we have kinetic and potential forms, which is linked to Rankine. On the other hand, we have the forms heat, electricity, magnetization, etc., which is connected to Mayer.

Regarding the former, Maxwell remarks that 'there may be other forms of energy, but we can only conceive of two, kinetic and potential'. This can be explained (Sect. 3.5). Kinetic is concerned with motion and potential subsumes rest. Now since

---

[89] "We cannot even assert that all energy must be either potential or kinetic, though we may not be able to conceive any other form" (ibid. p. 92).

[90] "it has been proved by experiment to be true within the limits of error of observation, in cases where the energy takes the forms of heat, magnetisation, electrification, &c." (ibid. p. 92).

[91] "the following statement is one which, if we cannot absolutely affirm its necessary truth, is worthy of being carefully tested, and traced into all the conclusions which are implied in it" (ibid. p. 92).

[92] "The reason for believing heat not to be a substance is that it can be generated, so that the quantity of it may be increased to any extent, and it can also be destroyed, though this operation requires certain conditions to be fulfilled" (ibid. p. 93).

[93] "The reason for believing heat to be a form of energy is that heat may be generated by the application of work, and that for every unit of heat which is generated a certain quantity of mechanical energy disappears. Besides, work may be done by the action of heat [...] Now when the appearance of one thing is strictly connected with the disappearance of another, so that [...] we conclude that the one has been formed at the expense of the other, and that they are both forms of the same thing. Hence we conclude that heat is energy in a peculiar form" (ibid. p. 93).

a particle can only be in motion or at rest, a third state is not imaginable. Therefore, there is no other form of energy.

The use of the concept of form in relation to heat, electricity, etc. leads to several forms of energy. Regarding heat and work, Maxwell points out that heat produces work and work gives rise to heat and this is the reason to believe that heat is a form of energy. Therefore, it is because heat appears in a production relationship which is interpreted as transformation of energy, that heat is taken as a form of energy.

As neither the details of the interpretations nor the names of the authors of the approaches used are indicated, the reader is not led to think that different theses are involved. As these theses appear as characteristics of energy, the concept of energy becomes confuse.

References

Carnot, S. (1824). *Réflexions sur la puissance motrice du feu.* Bachelier. (Rep. Paris: Éditions J. Gabay, 1990)

Clausius, R. (1850). Ueber die bewegende Kraft der Wärme und die Gesetze, welche sich daraus für die Wärmelehre selbst ableiten lassen. *Annalen Der Physik, 79*(368–397), 500–524.

Davy, H. (1839–1040). *Collected Works.* In: J. Davy (ed.) Smith, Elder and Co.

Joule, J. P. (1843). On the calorific effects of magneto-electricity, and on the mechanical value of heat. *The London, Edinburgh, and Dublin Philosophical Magazine and Journal of Science* Series 3. Vol. XXIII, pp. 263–276, 347–355, 435–443.

Joule, J. P. (1845). On the changes of temperature produced by the rarefaction and condensation of air. *Philosophical Magazine, 26*(3), 369–383.

Joule, J. P. (1847c) On matter, living force, and heat. Published in the Manchester 'Courier' newspaper, May 5 and 12. In Joule (1884, pp. 265–276).

Joule, J. P. (1850). On the mechanical equivalent of heat. *Philosophical Transactions of the RS of London, 140*, 61–82.

Joule, J. P. (1884) *The Scientific Papers of James Prescott Joule.* Vol. 1. The Physical Society. (Rep. London: Dawsons, 1963.)

Kragh, H. (2008). *Entropic creation: Religious contexts of thermodynamics and cosmology.* Ashgate.

Maxwell, J. (1873). *Theory of heat* (3rd ed.). Greenwood.

Neswald, E. (2006). *Thermodynamik als kultureller Kampfplatz: Zur Faszinationsgeschichte der Entropie, 1850–1915.* Rombach Verlag.

Rankine, W. (1850). Abstract of a paper on the hypothesis of molecular vortices, and its application to the mechanical theory of heat. *Proceedings of the Royal Society of Edinburgh, II*, 275–288.

Rankine, W. (1851). On the centrifugal theory of elasticity, as applied to gases and vapours. *Philosophical Magazine Series, 4*(2), 509–542.

Rankine, W. (1853). On the general law of the transformation of energy. *Philosophical Magazine, 34*, 106–117.

Rankine, W. (1855). Outlines of the science of energetics. *Edinburgh New Philosophical Journal, 2*, 120–141.

Thomson, W. (1848). On an absolute thermometric scale founded on Carnot's theory of the motive power of heat. *Philosophical Magazine, 33*, 313–317.

Thomson, W. (1849). An account of Carnot's theory of the motive power of heat; with numerical results deduced from Regnault's experiments of steam. *Transactions of the RS of Edinburgh, 16*, 541–574.

Thomson, W. (1851). On the dynamical theory of heat; with numerical results deduced from Mr Joule's equivalent of a thermal unit, and M. Regnault's observations on steam. *Transactions of the RS of Edinburgh, 20*, 261–298.
Thomson, W. (1851). On the dynamical theory of heat. On the quantities of mechanical energy contained in different states, as to temperature and density. *Transactions of the RS of Edinburgh, 20*, 475–482.
Thomson, W. (1852a). On a universal tendency in nature to the dissipation of mechanical energy. *Proceedings of the RS of Edinburgh, 3*, 139–142.
Thomson, W. (1852b). On the mechanical action of radiant heat or light. *Proceedings of the RS of Edinburgh, 3*, 108–110.
Thomson, W. (1854) On the mechanical antecedents of motion, heat, and light. In: Thomson (1884, pp. 34–40).
Thomson, W. (1884). *Mathematical and physical papers II*. Cambridge University Press.
Thomson, W., & Tait, P. (1862). Energy. *Good Words, 3*, 601–607.

Chapter 4
Reification of Energy

This chapter is devoted to a new phase of the concept: energy becomes a substance. Based on Maxwell's concept of the electromagnetic field, Poynting defends the thesis that energy moves in space. His argument is only theoretical. Nevertheless, Lodge finds here a reason to reinforce his idea of energy: it exists in bodies and in the aether and can be transferred between them. Understandably, energy was understood as a substance. There were, however, physicists who criticized the concept. Planck, Hertz and Poincaré will be considered. The third and final section of the chapter, "The Super Concept", approaches a generalization of the concept of energy never seen in a scientific concept. This was Ostwald's work.

4.1 Possession and Transfer of Energy

4.1.1 Lodge's Definition of Energy

In 1879, Lodge sent a letter to the editors of the *Philosophical Magazine*: Attempt at a systematic classification of the various forms of energy. He starts by criticizing the definition 'energy is the power to do work'.[1] The reason for the criticism is explained by the metaphor of capital: energy would have the power to perform work, as capital is purchasing power, but this power can only be realized if there are things to buy.[2] Lodge, therefore, criticizes the potentiality of the definition of energy (which came

[1] The "definition of energy […] "the power of doing work" […] seems a little unhappy" (Lodge, 1879, p. 279).

[2] "energy is power of doing work in precisely the same sense as capital is the power of buying goods. […] money is a power of buying goods. It does not, however, *necessarily* confer upon its owner any buying-power, because there may not be any accessible person to buy from; and if there be, he may have nothing to sell. Just so with energy: it usually […] confers upon the body possessing it acertain power of doing work, which power it loses when it has transferred it" (ibid. p. 279).

R. Lopes Coelho, *What Is Energy?*, History of Physics,
https://doi.org/10.1007/978-3-031-51855-3_4

from Thomson). To avoid this, he defines energy as the effect resulting from the work done on a body.[3] The magnitude of energy is naturally given by the amount of work done in producing this effect.[4]

Lodge arrives at the conservation of energy as follows. He starts from Newton's third law, arguing that every relationship between two bodies is of the action-reaction type. Now if work is done on a body, the body gains energy. If the body itself does work, then he says that the body has done anti-work, which is equivalent to loss of energy. Since work and anti-work are equal in absolute value, Lodge continues, energy is conserved. This conservation is expressed in the form "energy is neither produced nor destroyed but is simply transferred".[5] It is transferred from one body to another. What if there are no bodies between these two? In those cases, the aether works as a transfer medium, as we shall see next.

The forms of energy are justified by the variety of effects produced in the work done on bodies. Hence the types of energy will depend on the type of bodies. There are two main forms of energy:

- the free movement of bodies;
- tension between bodies.[6]

The former form is called kinetic and the latter, potential, although the author preferred other designations.[7] In the systematic classification of energy forms, 4 forms are indicated: translation, rotation, vibration, and tension. The table of forms of energy consists of 20 species, because Lodge considers 5 types of bodies. Four of these 5 types of bodies are systematized by size: planetary masses, ordinary masses or to our scale, particles or molecules, ultimate atoms. There is also "an unknown *something*, which is material enough to be capable of possessing energy, to disturbances in which electrical phenomena seem to be due, and of which probably an aspect has been called aether" (Lodge, 1879, p. 281). This entity will play an important role in the energy doctrine (Sect. 4.1.3).

Thomson, Rankine and Lodge

Thomson's mechanical energy had a potential character: it was the body's reserve of work available to humans, which could be used or not. Rankine expressed this

---

[3] "Whenever work is done *upon* a body, an effect is produced in it which is found to increase the working-power of that body (by an amount not greater than the work done); hence this effect is called *energy*" (ibid. pp. 278–9).

[4] "it is measured by the quantity of work done in producing it" (ibid. p. 279).

[5] "But in every action taking place between two bodies the work is equal to the antiwork (§ 3); hence the energy gained by the first body is equal to the energy lost by the second; or, on the whole, energy is neither produced nor destroyed, but is simply transferred from the second body to the first" (ibid. p. 279).

[6] "Energy [...] has two principal forms:—(1) The free motion of bodies relatively to one another; (2) The separation of bodies from one another against stress" (ibid. p. 280).

[7] "The energy possessed by matter in motion is called *Kinetic*. The energy possessed by matter exerting force is called *Potential*. It might with great propriety be called *Dynamic* energy; and it has been very conveniently called *Static* energy, in opposition to kinetic" (ibid. p. 281).

potentiality in the definition, energy is the ability of a body to do work. This capacity of the body is analogous to capital, which by itself does not do commerce. Therefore, Thomson's mechanical energy or Rankine's energy are designations given to something that is potential. Lodge intends, however, to link the concept of energy with actual (not potential) phenomena. Hence, Lodge's energy depends on the work done on the body. Regarding the measurement process, all three physicists agree that the measurement is to be made in units of work.

Conservation of the amount of work in a phenomenon is assumed by Thomson in 1851. Conservation of energy is admitted by Rankine in 1853. Lodge's conservation is related to Newton's 3rd law. This idea appears in the *Treatise of natural philosophy* by Thomson and Tait (1867, § 268) (Appendix L).

Concerning the concept of energy

Lodge's article brought about a change in the terms by which energy is conceptualized. 'Conservation' appears only once in the text: in Lodge's metaphor, capital is conserved.[8] The expression 'conservation of energy' appears in the summary of the paragraphs, which the author places at the end of the article.[9] 'Transformation' appears only once in the text, relating kinetic and potential energy.[10] The term that appears most often linked with energy is *transfer* or of the family. Another term linked to energy is *possession*, in the sense, bodies, matter and aether possess energy.

By these data, energy is to be understood by means of the concepts *possession* and *transfer*. Here is the origin of the concept: energy is something that is in bodies and is transferred from one body to another. If it is transferred from one body to another, it is in motion. Now, in bodies, it is not being transferred. Therefore, we have no reason to say that it is in motion. Thus, one might be led to think that it is at rest. This idea of energy at rest and in motion led to a further distinction between what is in the body, for which the term 'energy' was used, and what is in motion, such as heat or work. This appears in twentieth and twenty-first century textbooks (Chap. 5).

4.1.2 Poynting's Algorithm

In 1884, comes a paper by Poynting 'On the transfer of energy in the electromagnetic field' which will be pivotal to Lodge's thesis. The basic consideration is to look at a space where electric currents exist as a field of energy transformation, which would appear in the forms electric, magnetic, but also as work or heat.[11]

---

[8] "this is the law of the conservation of capital" (ibid. p. 279).

[9] "6. Conservation of energy, and first law of thermodynamics" (ibid. p. 285).

[10] "For instance, during every quarter-swing of a free pendulum, energy is being transformed from kinetic to potential, or *vice versâ*" (ibid. p. 285).

[11] "A SPACE containing electric currents may be regarded as a field where energy is transformed at certain points into the electric and magnetic kinds by means of batteries, dynamos, thermoelectric

Since according to Maxwell, electric currents essentially constitute a certain distribution of energy in and around the conductor, accompanied by transformation and movement of energy through the field, Poynting poses the question of how energy moves in the field.[12] The contribution of the article lies in the general law of energy transfer. By this law, the energy flows perpendicular to the flow of the electric and magnetic lines of force, and the amount of flow per unit area is a function of the electromotive intensity, the magnetic intensity, and the sine of the angle between the two.[13] After the presentation of the law, the article develops into applications, the first of which will be given as an example.

Electric current flows through a straight wire AB (Fig. 4.1). Before Maxwell, attention would have been focused on the conductor: energy was supposed to be carried by the current through the conductor.[14] Following Maxwell and thanks to the transfer law, continues Poynting, the explanation becomes the following. The lines of electric force have the direction of the current along the conductor; the lines of magnetic force are perpendicular to it and go around the conductor; by the law of energy transfer, energy flows from the outside to the inside.[15] The algorithm used leads to the result i^2R, where i stands for the current intensity and R the resistance. Now i^2R is Joule's

actions, and so on, while in other parts of the field this energy is again transformed into heat, work done by electromagnetic forces, or any form of energy yielded by currents" (Poynting, 1884, p. 343).

[12] "According to MAXWELL's theory, currents consist essentially in a certain distribution of energy in and around a conductor, accompanied by transformation and consequent movement of energy through the field" (ibid. p. 343). "The aim of this paper is to prove that there is a general law for the transfer of energy, according to which it moves at any point perpendicularly to the plane containing the lines of electric force and magnetic force [...]" (ibid. p. 344).

[13] "On interpreting the expression it is found that it implies that the energy flows as stated before, that is, perpendicularly to the plane containing the lines of electric and magnetic force, that the amount crossing unit area per second of this plane is equal to the product

$$\frac{\textit{electromotive intensity} \times \textit{magnetic intensity} \times \textit{sine included angle}}{4\pi}$$

while the direction of flow [...]" (ibid. p. 345).

[14] "Formerly a current was regarded as something travelling along a conductor, attention being chiefly directed to the conductor, and the energy which appeared at any part of the circuit, if considered at all, was supposed to be conveyed thither through the conductor by the current" (ibid. p. 343).

[15] "In this case very near the wire, and within it, the lines of magnetic force are circles round the axis of the wire. The lines of electric force are along the wire [...] energy is therefore flowing in perpendicularly through the surface, that is, along the radius towards the axis" (ibid. p. 350).

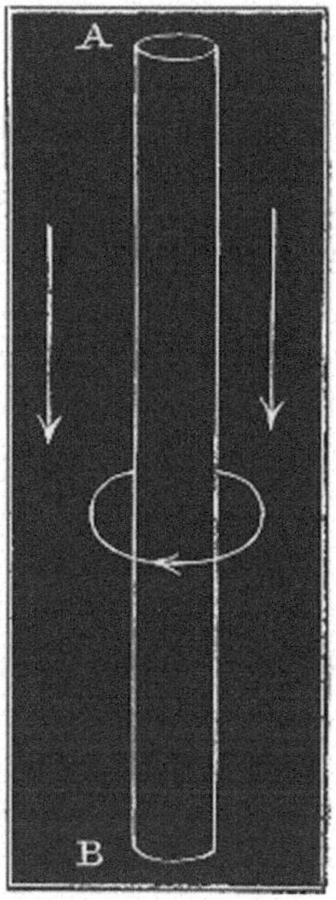

Fig. 4.1 The lines of electric force and lines of magnetic force (Poynting, 1884, p. 282)

expression for the heat developed.[16] Poynting's interpretation is then the following: the surrounding energy flows into the conductor and is transformed there into heat.[17]

Conclusion

The article deals with the transfer of energy, in the context of which expressions such as flow, energy travel, or analogues arise. There are also expressions such as distribution of energy around the conductor, contained or resident energy. In these

[16] "Let r be the radius of the wire, i the current along it, α the magnetic intensity at the surface, P the electromotive intensity at any point within the wire, and V the difference of potential between the two ends. Then the area of a length l of the wire is 2πrl, and the energy entering from the outside per second is

$$\frac{area \times E.M.I \times M.I}{4\pi} = \frac{2\pi rl.P.\alpha}{4\pi} = \frac{2\pi r\alpha.Pl}{4\pi} = \frac{4\pi iV}{4\pi} = iV$$

for the line integral of the magnetic intensity $2\pi r\alpha$ round the wire is $4\pi x$ current through it, and Pl=V.

But by OHM's law V = iR and iV = i^2R, or the heat developed according to JOULE's law" (ibid. pp. 350–1).

[17] "It seems then that none of the energy of a current travels along the wire, but that it comes in from the nonconducting medium surrounding the wire, that as soon as it enters it begins to be transformed into heat, the amount crossing successive layers of the wire decreasing till by the time the centre is reached, where there is no magnetic force, and therefore no energy passing, it has all been transformed into heat. A conduction current then may be said to consist of this inward flow of energy with its accompanying magnetic and electromotive forces, and the transformation of the energy into heat within the conductor" (ibid. p. 351).

cases, a place of permanence of energy is indicated; in the previous ones, movements of energy. The result of both is that energy is something that is in one place and moves from one place to another.

The idea is developed from the mathematical expression of the so-called law of energy transfer. The calculation leads to a vector. This is what Poynting relates to energy. The trajectory and motions of energy therefore have their origin in calculus. However, the calculus is not in itself a physical argument. Poynting also expressed reservations: experiments would have to be performed to prove if there is movement of energy. He is aware that his explanation is based on a "mental image". He also states that it is necessary to accept Maxwell's theory. For the author himself, this would therefore be theoretical work, to be complemented by experimental work.

4.1.3 Lodge's Identity of Energy

'On the identity of energy' is the title of Lodge's (1885) paper, whose leitmotif was Poynting's paper. According to Lodge, Poynting introduced the idea of continuity into the existence of energy.[18] Now, he adds, one can follow an amount of energy to its emergency in another place and in another form.[19] Herein lies the identity of energy.[20] This idea, Lodge continues, represents an extension of the principle of conservation because the principle was based on the constancy of the quantity, whereas now we know the trajectory itself.The idea of an energy trajectory leads Lodge to focus on the means of energy transfer, as we will see.

The new doctrine of conservation of energy, as he calls it, is based on two premises that support or explain that transfer: force is a component of an action-reaction relationship; and this relationship is always by contact.[21] Here is how this is justified. If a body A performs work on B, B exerts a reaction, following Newton's third law,

---

[18] "In that paper he introduces the idea of continuity in the existence of energy [...] whenever energy is transferred from one place to another at a distance, it is not to be regarded as destroyed at one place and recreated at another, but it is to be regarded as transferred, just as so much matter would have to be transferred; and accordingly we may seek for it in the intervening space, and may study the paths by which it travels" (Lodge 1885, p. 482).

[19] "The conservation of energy was satisfied by the *total quantity* remaining unaltered; there was no individuality about it: one form might die out, provided another form simultaneously appeared elsewhere in equal quantity. On the new plan we may label a bit of energy and trace its motion and change of form, just as we may ticket a piece of matter so as to identify it in other places under other conditions" (ibid. p. 482).

[20] "The energy may be watched at every instant. Its existence is continuous; it possesses identity" (ibid. p. 483).

[21] "the doctrine may be proved rigidly and instantaneously from two very simple premises, viz. Newton's law of motion on the one hand, and the denial of action at a distance on the other" (ibid. p. 482).

and the connection between A and B is the means of energy transfer.[22] If the two bodies are not actually in contact, Lodge admits 'a something' filling the space, which he presumes to be the aether, in which energy moves.[23] In this way Lodge achieves contact between bodies in any situation and thus has a channel for the transfer of energy.

One advantage of the new doctrine, according to Lodge, is to solve a typical difficulty of the previous theory, potential energy.[24] If a stone falls, the potential energy is not to be thought of in the grave nor in the earth, but rather in the surrounding medium. It is this medium, he explains, that presses the stone and the earth toward each other.[25] What was then said of the fall of a grave (energy gradually changes from potential to kinetic while remaining in the stone) is for the author meaningless.[26] The pendulum motion is another example. The potential energy of the medium is transferred to the moving bob, appears there as kinetic energy and is transferred back to the medium as potential energy.[27]

The falling stone and the pendulum motion are the examples given of the transformation and transfer of the two fundamental forms of energy, kinetic and potential. The justification of these two forms is given as follows. Work consists of the product of two factors, force and motion.[28] The two forms of energy correspond to each of the factors: potential corresponds to force and kinetic corresponds to motion.[29] More exactly, potential energy corresponds to force combined with elasticity and kinetic

---

[22] "The stress between A and B is the means of transferring energy from A to B, directly motion takes place in the sense AB. And the energy cannot *jump* from A to B, it is transferred across their point of contact" (ibid. p. 483).

[23] "A may be a molecule of matter, M may be the nearest molecule to it, and energy may be transferred from A to M, but not directly; A cannot act on M, cannot do work on it, because of the intervening gap. A can act on B, transferring its energy to B, B can act on C, C on D [...] What B, C, D, ... L are, I do not presume to say; but of course one supposes them to be successive portions of the perfectly continuous space-filling medium Aether" (ibid. p. 484).

[24] "In the older and more hazy view of conservation of energy the idea of "potential energy" has always been felt to be a difficulty [...] it was not easy or possible always to form a clear and consistent mental image of what was physically meant by it [...] The usual ideas and language current about potential energy are proper to notions of action at a distance" (ibid. p. 484).

[25] "When universal contact action is admitted, the haze disappears; the energy is seen to be possessed, not by stone or by earth or by both of them, but by the medium which surrounds both and presses them together" (ibid. p. 484).

[26] "the common mode of treating a falling weight, saying that its energy gradually transforms itself from potential to kinetic but remains in the stone all the time, is, strictly speaking, nonsense. The fact is the stone never had any potential energy, no rigid body can have any; the gravitation medium had it however, and kept on transferring it to the stone all the time it was descending" (ibid. p. 486).

[27] "A pendulum exhibits the alternation of energy from the kinetic to the potential form and the accompanying transfer from matter to medium, at every half-swing" (ibid. p. 486).

[28] "Energy has two fundamental forms because work has two factors, force and motion, F, *s*" (ibid. p. 484).

[29] "The two forms of energy correspond to the factors in the product work.
"Potential" energy correspond to F.
"Kinetic" energy correspond to *s*" (ibid. p. 485).

energy, to motion combined with inertia.[30] So, for example, a pillar supporting a structure exerts force, but has no potential energy; unlike an arrow bow, which not only exerts force, but because it has elasticity can also propel the arrow.[31] This applies *mutatis mutandis* to motion. It is not enough for the body to be in motion. It must be able to produce work, which is attributed to the inertia of bodies.[32] Both forms of energy are seen as potential work, requiring the other element to do work.[33]

Conclusion

Lodge's contribution in this paper concerns the concept of energy. The terminology used is close to that of 1879. The term *transfer* or of the family usually appears linked to energy. Transfer and transformation are linked by the theory: whenever there is a transfer of energy, there is a change of form. *Possession* or of the family are terms expressing the relation of bodies or the medium to energy. *The energy existing in the medium distinguishes the new doctrine* from the previous one.

The advantage of the new doctrine highlighted by Lodge is the explanation of potential energy. This concept was created by Rankine. The stone at a certain height, the example given by Lodge, has potential energy. This energy is given by the weight of the stone and the distance it is at. The weight of the body depends on the gravitational action of the earth. Then the question was asked about where the potential energy is: in the body, in the Earth, or in both. Lodge answers, the potential energy is in the medium. This is what presses the body and the Earth towards each other. That question—where is the energy located—as well as Lodge's answer presuppose that energy is substantial because being substantial is a necessary condition for it to be able to reside in a body, in the earth, in the medium. Therefore, the advantage pointed out only holds if one admits energy as a substance.[34]

Energy in motion in space must be substantial. This idea of the motion of energy in space is based on the law of energy transfer in the electromagnetic field.[35] It turns

---

30 "Kinetic energy corresponds to motion combined with inertia, so that the motion shall continue even against some force [...] Potential energy corresponds to force combined with elasticity (or something like it)" (ibid. p. 485).

31 "A strained bow is exerting force and possesses energy. A pillar supporting a roof is exerting force, but possesses no energy [...] Thus, then, for a body to possess potential energy we must have two things—the exertion of a force, together with a guarantee that that force shall be exerted over a certain distance; *i.e.* a continuance of the force even after motion is permitted" (ibid. p. 485).

32 ""A body in motion possesses energy;" but is it so necessarily? [...] Suppose it stops the instant you give it work to do—the instant you make it exert force. It is evident you must have not merely motion, you must have a guarantee of *persistence* of motion, the body must possess inertia" (ibid. p. 485).

33 "Both forms of energy are *potential work* [...] Kinetic energy requires the Force factor to do work. Potential energy requires the Motion factor to do work" (ibid. p. 485).

34 In the same year, Tait defended the same idea: "In the physical universe there are but two classses of things, Matter and Energy" (Tait, 1885, p. 2).

35 In the *Treatise*, Maxwell wrote: "When light is emitted, a certain amount of energy is expended by the luminous body, and if the light is absorbed by another body, this body becomes heated [...] During the interval of time after the light left the first body and before it reached the second, it must have existed as energy in the intervening space." (Maxwell, 1873, § 782).

out, however, that according to Poynting, experimental evidence of the trajectories of energy would be lacking. Lodge did not carry out any experimental work. Therefore, what Poynting pointed out was missing, was still missing.

4.2 Conceptual Difficulties and Criticism

4.2.1 Planck

In the middle of the 1880s, the Göttingen Philosophical Faculty denoted some confusion about the concept of energy and the principle of energy conservation. Indeed, it poses the following question for competition: since Thomas Young, many physicists have attributed energy to bodies and, since William Thomson, there has been talk of a principle of conservation of energy valid for all bodies, by which one seems to understand what Helmholtz had expressed as the principle of conservation of force. Once the question was formulated, the following was required: a historical study of the semantic evolution of energy and its use in physics; an investigation into the forms of energy; and how the principle of conservation of energy can be formulated and proved as a general law of nature.[36]

Planck competed in this award. The submitted work appears in book form in 1887, with the title *The Principle of Conservation of Energy*. The first part of the book is devoted to the historical development of the concept; the second, to the definition of energy and proof of the principle of conservation; and the third and final part, to the forms of energy: mechanical, thermal, chemical, electrical, and magnetic. The second part of the work provides the main object of the present study.

Planck presents the following definition of energy:

> "we call energy (capacity to do work) of a system in a given state *the value of all the effects, measured in units of work, which originate outside the system, when it passes from that state to another fixed arbitrarily as the null state, whatever the mode of passage*".[37]

In this definition, two parts can be considered: one concerns the meaning of energy and the other, the magnitude. By energy is understood the capacity of doing work. The

[36] "Seit Thomas Young [...] wird den Körpern von vielen Physikern *Energie* zugeschrieben, und seit William Thomson [...] wird häufig das *Prinzip der Erhaltung der Energie* als ein für alle Körper gültiges ausgesprochen, worunter dasselbe Prinzip verstanden zu werden scheint, was schon früher von Helmholtz unter dem Namen des *Prinzips der Erhaltung der Kraft* ausgesprochen war. Es wird nun zunächst eine genaue *historische* Entwicklung der Bedeutung und des Gebrauchs des Wortes *Energie in der Physik* verlangt; sodann eine *gründliche physikalische* Untersuchung, ob *verschiedene Arten* der Energie zu unterscheiden [...] sei; endlich in welcher Weise das *Prinzip der Erhaltung der Energie* als allgemein *gültiges Naturgesetz* aufgestellt und bewiesen werden könne" (Planck, 1921, Preface to the first edition).

[37] "bezeichnen wir die Energie (Fähigkeit, Arbeit zu leisten) eines materiellen Systems in einem bestimmten Zustand als den in mechanischen Arbeitseinheiten gemessenen Betrag aller Wirkungen, welche außerhalb des Systems hervorgebracht werden, wenn dasselbe aus seinem Zustand auf beliebige Weise in einen nach Willkür fixierten Nullzustand übergeht" (ibid. p. 104).

magnitude is given by the value of the external effects of the system in units of work, when the system passes from a given state to the standard state. To give the value of the external effects in units of work, we need to have the mechanical equivalents of these effects.[38] If the mechanical equivalent of the system effects in a given passage is unknown, we cannot account for this passage in terms of work.[39] Planck suggests overcoming this difficulty in two ways. One is to recreate that passage and use this effect to produce work. The other is to produce that effect through work.[40] If neither of these ways is feasible, the definition is useless, says Planck.[41]

Those hypotheses assume that using work to produce the magnitude of which we do not know the mechanical equivalent is the same as using that magnitude to produce work. Here Planck struggles with the question of whether or not the amount of work depends on the process of transformation.[42] To overcome this difficulty, Planck resorts to the principle of conservation of energy in the following form: the mechanical equivalent of the passage of the system from one state to the standard state does not depend on the mode of passage.[43] Thus, the final part of the definition

[38] "Unter den "außerhalb des Systems hervorgebrachten Wirkungen" oder kürzer: unter den „äußeren Wirkungen" wollen wir alle am Schluß des Prozesses in der Natur eingetretenen Veränderungen verstehen, welche mit der Lage und Beschaffenheit der umgebenden, nicht in das System einbegriffenen Körper zusammenhängen, darunter also z. B. auch die Veränderung der Lage des Systems relativ zur Umgebung" (p. 104). "Was ferner den in der Definition gebrauchten Ausdruck: "Der in mechanischen Arbeitseinheiten gemessene Betrag" (kurz: *Arbeitswert, mechanisches Äquivalent*) der äußeren Wirkungen betrifft, so hat derselbe natürlich nur unter der Voraussetzung einen bestimmten Sinn, daß entweder die äußeren Wirkungen an sich lediglich mechanischer Natur sind [...] oder, falls sie von irgend anderer Art sind, daß dann ihr mechanisches Äquivalent schon anderweitig bekannt ist" (ibid. p. 105).

[39] "nehmen wir z. B. an, die äußeren Wirkungen beständen in der Erzeugung irgend einer eigentümlichen Veränderung, etwa eines gewissen Agens, dessen Arbeitswert unbekannt sei, so läßt natürlicherweise die Definition zunächst im Stich" (ibid. p. 105).

[40] "man muß sich dadurch zu helfen suchen, daß man das neu erzeugte Agens auf irgend eine Weise wieder fortschafft, indem man es etwa zur Leistung mechanischer Arbeit oder zur Hervorbringung solcher Wirkungen verbraucht, die auf mechanisches Arbeitsmaß reduzibel sind [...] dann stellt sich das mechanische Äquivalent einer Wirkung als diejenige Arbeitsmenge dar, *in welche sich diese Wirkung verwandeln läßt*" (ibid. p. 105).

[41] "Es ist aber auch sehr wohl der Fall denkbar, daß es überhaupt unmöglich ist, das neue Agens ganz in mechanische Wirkungen zu verwandeln, und in diesem Falle wird die für den Begriff des Arbeitswertes gegebene Erklärung, also auch die Definition der Energie, hinfällig" (ibid. pp. 105–6).

[42] "Hierbei bleibt es übrigens noch ganz dahingestellt, ob die Arbeitsmenge verschieden ausfällt, wenn die Verwandlung auf verschiedene Weise vorgenommen wird" (ibid. p. 105).

[43] "*Die Energie eines materiellen Systems in einem bestimmten Zustand, genommen in bezug auf einen bestimmten anderen Zustand als Nullzustand, hat einen eindeutigen Wert*, oder mit anderen Worten [...] Der in mechanischen Arbeitseinheiten gemessene Betrag (das mechanische Äquivalent, der Arbeitswert) aller Wirkungen [...] hat einen eindeutigen Wert, ist also unabhängig von der Art des Überganges" (ibid. pp. 110–1).

of energy, "whatever the mode of passage", becomes valid.[44] It is from this formulation of the principle that, according to Planck, the definition of energy becomes complete.[45]

From the principle, Planck draws a set of consequences. Some are exposed through a kind of algebra of passages of the system from one state to another. Among these consequences arises a second formulation of the principle of conservation of energy. Let us consider some of the consequences and the second formulation of the principle.

Let A, B, C, D, ... M, N be states of the system and A–B, B–C, etc. the processes that take the system from state A to B, from B to C, etc., represented by [AB], [BC], etc.. Planck first proposition states that the energy of state A referred to state N, taken as the null state, is equal to the sum of the intermediate energies, i.e., [AN] = [AB] + [BC] + ... + [MN].[46]

The additive constant that appears in the energy equation is expressed in the following terms. If the energy of a given state is referred to as the N state, the process will be given as [AN]. If the standard state is N', the process will be given as [AN']. Consequently, there is a difference, [AN]-[AN'] = [N'N], which accounts for the additive constant of the equation.[47] There is something that depends on the choice of the standard state and not on the phenomenon.

An especially interesting consequence is the one where [AB] = 0. Taking N as the standard state, one has [AN] = [BN].[48] This means that in the passage [AB] no external work was performed. This is connected to the second formulation of the conservation principle: the energy of the system without external action is conserved.[49]

---

[44] "In der Tat können wir in jedem Falle, wo aus irgend welchen Gründen unsere Definition nicht zum Ziele führt, uns immer noch auf eine andere Weise helfen, nämlich dadurch, daß wir den betreffenden Fall vorläufig ganz von der Betrachtung ausschließen und die für ihn zugebende Definition des Energiebegriffes erst bei einer späteren Gelegenheit (S.113) nachholen, wo wir im Besitze verschiedener Sätze sein werden, welche die Berechnung des Wertes der Energie unter allen Umständen gestatten. Zu diesen Sätzen gelangen wir durch die Aufstellung des *Prinzips der Erhaltung der Energie*" (ibid. p. 110).

[45] "Wir können nun die hier abgeleiteten Sätze zugleich benutzen, um die allgemeine *Definition* des Energiebegriffs dadurch zu vervollständigen, daß wir sie auch auf diejenigen Fälle ausdehnen, [...]" (ibid. p. 113).

[46] "Die Energie des Systems im Zustand A, bezogen auf den Nullzustand N ist gleich der Summe der Energien in den Zuständen A, B, C, ..., M, bezogen auf die respektiven Nullzustände: B, C, D, ..., N" (ibid. p. 112).

[47] "Wenn wir die Energie eines materiellen Systems in einem bestimmten Zustand A einmal auf den Zustand N, dann auf einen anderen Zustand N' als Nullzustand beziehen, so folgt aus der Relation: $[AN] - [AN'] = [AN] + [N'A] = [N'N]$ [...] Lassen wir daher bei der Bestimmung der Energie die Wahl des Nullzustandes ganz offen, so wird in dem Ausdruck der Energie nur eine gewisse additive Konstante unbestimmt gelassen" (ibid. p. 114).

[48] "Wenn speziell der Prozeß in der Weise vorsich geht, daß in der äußeren Umgebung gar keine Wirkungen stattfinden, dann ist [AB] = 0, also [AN] = [BN]: die Energie im Zustand A ist gleich der im Zustand B" (ibid. p. 115).

[49] "Die Energie eines materiellen Systems, in bezug auf einen beliebigen Nullzustand, ändert sich also nicht, wenn bei Ausführung irgend eines Prozesses keine äußere Veränderung eintritt, oder mit anderen Worten: wenn in dem System nur innere Wirkungen stattfinden. In dieser Form stellt

This formulation of the principle leads to a conceptual question. Planck asks himself about how to think of the energy of the system in which energy is conserved.[50] He proposes to imagine energy as a kind of reserve, a 'capital'.[51] This conception would be very practical and of easy intuition, by virtue of the analogy with matter. Just as the sum of the total mass of a body is equal to the sum of the masses of its parts, the total energy of the system would be equal to the sum of the various forms of energy.[52]

Although Planck sees the advantages of the analogy between matter and energy and sees therein the reason for the rapid development of the doctrine,[53] the question is raised of whether it is legitimate to think of energy as a substance.[54] There is at once a difficulty with this concept of energy. According to Planck, one cannot indicate a place for energy in a system.[55] He then admits that the material concept of energy will one day be superseded. At that time, he justifies it by its heuristic function, the search for new forms of energy[56] and its intuitive character.[57]

sich das Prinzip als das der Erhaltung der Energie dar, und diese Form ist es nun, die durch eine etwas veränderte Auffassung des Begriffes der Energie sich so ungemein bequem für die direkte Anschauung und fruchtbar für die weitere Behandlung erweist" (ibid. p. 115).

[50] "Dieser Satz führt uns dazu, die in einem System enthaltene Energie als eine begrifflich von den äußeren Wirkungen unabhängig bestehende Größe aufzufassen" (ibid. p. 116).

[51] "Nun haben wir uns die Energie als im System selbst befindlich vorzustellen, als eine Art *Vorrat* (nach C. Neumann: "Kapital"), welcher durch innere Wirkungen unzerstörbar ist" (ibid. p. 116).

[52] "diese Auffassung ist für die unmittelbare Anschauung überaus bequem durch ihre Analogie mit dem Verhalten der Materie, die auch in verschiedene Formen überführbar, aber nach ihrer Quantität (Masse) unveränderlich ist. Ebenso wie die Gesamtmasse eines Körpers sich als die Summe der Massen der einzelnen in demselben enthaltenen chemischen Substanzen darstellt, so setzt sich die Energie eines Systems zusammen aus der Summation der einzelnen Energiearten" (ibid. p. 116).

[53] "Ohne Zweifel beruht zum großen Teil auf dieser Analogie die verhältnismäßig überraschende Leichtigkeit und die sieghafte Klarheit, mit der sich das Prinzip der Erhaltung der Energie binnen weniger Jahre die allgemeine Anerkennung eroberte und in der Überzeugung eines jeden festsetzte" (ibid. p. 116).

[54] "Man könnte hier die Frage aufwerfen, ob es denn wirklich für die gesunde Weiterentwicklung des Prinzips von Nutzen ist, in dieser Weise von der primären Definition des Begriffes abzuweichen und ihm eine spezielle substanzielle Deutung zu geben" (ibid. pp. 116–7).

[55] "die Unbestimmtheit liegt dann im *Begriff* der Energie, man kennt den Platz nicht, den man ihr anweisen soll, und hat auch kein Mittel, ihn zu finden" (ibid. p. 117).

[56] "[...] ist [...] unverkennbar, daß mit der hier in Rede stehenden substanziellen Deutung des Begriffes der Energie nicht nur eine Vermehrung der Anschaulichkeit, sondern auch ein direkter Fortschritt in der Erkenntnis verbunden ist. Dieser Fortschritt beruht auf der Anregung zur weiteren physikalischen Forschung. Man wird sich nun nicht mehr damit begnügen, den Zahlenwert der Energie des Systems zu kennen, sondern man wird versuchen, die Existenz der verschiedenen Arten der Energie an den verschiedenen Elementen des Systems im Einzelnen nachzuweisen, und den Übergang in andere Formen und zu anderen Elementen ebenso verfolgen, wie die Bewegung eines Quantums Materie im Raum" (ibid. pp. 117–8).

[57] "Gewiß ist zuzugeben, daß diese (sozusagen materielle) Auffassung der Energie als eines Vorrats von Wirkungen, dessen Menge durch den augenblicklichen Zustand des materiellen Systems bestimmt ist, möglicherweise später einmal ihre Dienste getan haben und einer anderen, allgemeineren und höheren, Vorstellung Platz machen wird: gegenwärtig ist es jedenfalls Sache der

Conclusion

Planck wants to have a definition of energy, in which energy corresponds to something in phenomena. The definition does include a method of measurement but a difficulty arises with its application: one cannot always measure the mechanical equivalent. To overcome this obstacle, Planck resorts to the principle of conservation. However, in this way the definition is jeopardized: it needs a postulate; therefore, if the latter is false, the definition is useless. With the second formulation of the principle, that the energy of an isolated system does not vary, another difficulty arises, that of understanding energy. The need for a concept of energy capable of subsuming the idea that energy exists in the system leads one to think of it as a substance. If energy is a kind of substance, it should be possible to localize it in the system. Experiments, however, do not allow us to localize it. For this reason, Planck predicts that that concept of energy will one day be superseded.

4.2.2 *Hertz's Criticism*

In 1894, Heinrich Hertz's *Principles of Mechanics* appeared. At the time, mechanics was the foundation of the whole of physics. Therefore, the foundation of mechanics itself, which was Hertz's focus, was an important issue. He systematizes the mechanics of that time into two theories, which he calls *images*. The first image is based on Newton's laws and d'Alembert's principle and the second image concerns the theory of energy. This is characterized by four primitive concepts—space, time, mass and energy—and one axiom—Hamilton's principle. The concept of energy constitutes an obstacle to the acceptance of this theory.

Energetics is not mature enough to provide a definition of energy, says Hertz.[58] According to the use given to the concept, however, he assumes that energy is conceived as a substance.[59] A first difficulty arises from the fact that the possible substance comes in such different forms as kinetic and potential energy.[60] Another difficulty concerns the latter, because according to Hertz potential energy contradicts the character of substance[61]:

---

physikalischen Forschung, diese Auffassung als die anschaulichste und fruchtbarste überall bis ins einzelne durchzubilden und ihre Konsequenzen an der Hand der Erfahrung zu prüfen" (ibid. p. 118).

[58] "zu einem befriedigenden und abschließenden Ergebnis scheint diese ganze Anschauungsweise noch nicht gelangt" (Hertz, 1894, p. 26).

[59] "Mehrere ausgezeichnete Physiker versuchen heutzutage, der Energie so sehr die Eigenschaften der Substanz zu leihen, daß sie annehmen, jede kleinste Menge derselben sei zu jeder Zeit an einen bestimmten Ort des Raumes geknüpft und bewahre bei allem Wechsel desselben und bei aller Verwandlung der Energie in neue Formen dennoch ihre Identität" (ibid. pp. 25–6).

[60] "Eine besondere Schwierigkeit muß auch von vornherein der Umstand bereiten, daß die angeblich substanzartige Energie in zwei so gänzlich verschiedenen Formen auftritt, wie es die kinetische und die potentielle Form sind" (ibid. p. 26).

[61] "die potentielle Energie [...], welche eine selbständige Feststellung fordert, widerstrebt zugleich jeder Definition, welche ihr die Eigenschaften einer Substanz beilegt" (ibid. p. 26).

- the quantity of a substance is necessarily a positive quantity, whereas the potential energy of a system can be negative[62];
- in the analytical expression of the quantity of a substance, an additive constant has as much significance as the rest, whereas the additive constant in the expression of potential energy is not significant[63];
- the content of substance in a given system depends on the state of the system, whereas the potential energy of a system depends on the existence of distant masses, which probably, Hertz adds, never had any influence on the system.[64]

Conclusion

The difficulty with energy lies in the following: it is conceived as a substance when it contradicts the very concept of substance. In conformity with Hertz's criteria of evaluating a physical theory, energetics is logically non-permissible.

4.2.3 Poincaré's Criticism

Three years after the *Principles of Mechanics*, Poincaré wrote an article on this book. He corroborates the advantages of a mechanical theory based on the concept of energy over a theory based on force pointed out by Hertz: dispensing with atoms; and being less incomplete, by eliminating unnatural motions consistent with the classical theory.[65] Hertz's objections to the concept of energy are seen as being of a "quasi-metaphysical" order.[66] However, Poincaré corroborates the difficulty in locating energy, for if the kinetic energy could be located in the mobile, it would no longer be clear where to locate the potential.[67] He also refers to the arbitrary constant

---

[62] "Die Menge einer Substanz ist eine notwendig positive Größe; die in einem System enthaltene potentielle Energie scheuen wir uns nicht, als negativ anzunehmen" (ibid. p. 26).

[63] "Bedeutet ein analytischer Ausdruck die Menge einer Substanz, so hat eine additive Konstante in dem Ausdruck dieselbe Wichtigkeit wie der Rest; in dem Ausdruck für die potentielle Energie eines Systems hat die additive Konstante niemals eine Bedeutung" (ibid. p. 26).

[64] "Endlich kann der Inhalt eines physikalischen Systems an einer Substanz nur abhängen von dem Zustande des Systems selbst; der Inhalt gegebener Materie an potentieller Energie aber hängt ab von dem Vorhandensein entfernter Massen, welche vielleicht niemals Einfluß auf das System hatten" (ibid. p. 26).

[65] "La théorie énergétique présente sur la théorie classique les avantages suivants:

1° Elle est moins incomplète; c'est-à-dire que les principes de la conservation de l'énergie et de Hamilton nous apprennent plus que les principes fondamentaux de la théorie classique et excluent certains mouvements que la Nature ne réalise pas et qui seraient compatibles avec la théorie classique;

2° Elle nous dispense de l'hypothèse des atomes, qu'il était presque impossible d'éviter avec la théorie classique" (Poincaré, 1897, p. 738).

[66] "Ce sont d'autres objections, d'ordre presque métaphysique, que Hertz développe le plus longuement" (ibid. p. 739).

[67] "D'autre part, pour *matérialiser* l'énergie, il faut la *localiser*; pour l'énergie cinétique, cela est facile, mais il n'en est pas de même pour l'énergie potentielle. Où localiser l'énergie potentielle due

in the energy equation as something that 'shocks the spirit'[68] and shows, through an example, that energy can be negative.[69]

In addition to his acceptance of Hertz's objections, Poincaré has his own objection to the energy theory, which is the following. If there is a system of points, whose forces depend on distances, then there is a certain quantity accessible to experiment that remains constant. This quantity is given by the sum of two terms, continues Poincaré, where one depends only on the *position* of the points (potential energy) and the other is proportional to the square of the *velocities* (kinetic energy). The difficulty arises if the forces involved in the potential energy, also depend on velocity.

Poincaré gives as an example of such a force, the one suggested by Weber: the action between two electric molecules also depends on velocity. If potential energy does not depend only on position but also on velocity, then it is not possible to sort out what depends on velocities or is independent of them.

If one considers several forms of energy, Poincaré continues, the principle takes the form

$$T + U + Q = constant \tag{4.3}$$

where T represents the kinetic energy, U the potential energy of position, and Q the molecular internal energy, of chemical or electrical form. There is no objection, continues Poincaré, if the decomposition can be performed: T is proportional to the squares of the velocities; U does not depend on the velocities; and Q depends only on the internal state. Electrified bodies provide, however, cases where this sorting is not possible. In such cases, says Poincaré, the principle of conservation of energy tells us that something remains constant, i.e., there is something that is constant, but this something has not the property of decreasing in one component and increasing in the others in due proportion. Thus, Poincaré concludes, in that way, the principle is a tautology because if there are laws of nature, something must remain constant.[70]

4.3 The Super Concept

In 1908, Ostwald published *The Energy*. In this book, he presents the energetic interpretation of physical and chemical phenomena, but also of life, of the phenomena of the spirit and of life in society.

à l'attraction de deux astres? Est-ce dans l'un des deux astres? Est-ce dans les deux? Est-ce dans le milieu intermédiaire?" (ibid. p. 739).

[68] "Comme nous disposons de la constante *c*, nous pouvons la supposer assez grande pour que l'énergie soit positive; il y a dejà là quelque chose d'arbitraire qui choque l'esprit" (ibid. p. 739).

[69] "Si l'énergie est pour ainsi dire *matérialisée*, elle devra ètre toujours positive. [...] nous ne pouvons pas assurer que l'énergie demeurera toujours positive" (ibid. p. 739).

[70] "Il ne nous reste plus qu'un énoncé pour le principe de la conservation de l'énergie; *il y a quelque chose qui demeure constant*. Sous cette forme, il se trouve à son tour hors des atteintes de l'expérience et se réduit à une sorte de tautologie" (ibid. p. 739).

In the introduction to the book, one reads, ‘energy embodies the real’. This is explained as follows: energy is real because it is active in what happens; and it is real because it constitutes the content of events.[71] This has consequences. Ostwald explicitly claims that energy makes the concept of matter superfluous. The reason for this is that the properties attributed to matter can be given by the forms of energy. Thus, for example, bodies are said to have mass and weight. Now mass corresponds to one of the factors of the energy of motion and weight is a factor of the energy of gravitation. (The energy of motion is understood as the product of mass by the square of velocity[72] and the energy of gravitation as the product of weight by height.[73]) Ostwald calls them material factors, simply because they are related to what is understood as matter.[74] Mass and weight are in this context only examples. Energy of form,[75] energy of volume,[76] of surface, of distance and with space as the field of activity of energies[77] provide Ostwald with an energetic expression for what was ascribed to matter. Hence, he goes on to dispense with this concept.[78] In addition, he adds, the concept of energy is able to account for properties of bodies, such as heat or chemical properties, which for one reason or another are not part of the concept of matter.[79] In sum, the concept of

[71] “Die Energie ist daher in allen realen oder konkreten Dingen als wesentlicher Bestandteil enthalten, der niemals fehlt, und insofern können wir sagen, *daß in der Energie sich das eigentlich Reale verkörpert.*

Und zwar ist die Energie das Wirkliche in zweierlei Sinn. Sie ist das Wirkliche insofern, als sie das *Wirkende* ist; wo irgend etwas geschieht, kann man auch den Grund dieses Geschehens durch Kennzeichnung der beteiligten Energien angeben. Und zweitens ist sie das Wirkliche insofern, als sie den *Inhalt* des Geschehens anzugeben gestattet” (Ostwald, 1912a, p. 5).

[72] “Nehmen wir als Beispiele die *Masse*, den Extensitätsfaktor der Bewegungsenergie” (ibid. p. 112). “ist die Masse nicht etwa umgekehrt proportional der Geschwindigkeit zu setzen, sondern umgekehrt proportional dem *Quadrat* der Geschwindigkeit” (ibid. p. 119).

[73] “Es gibt also eine Schwereenergie oder Gravitationsenergie, und ihre Faktoren lassen sich leicht erkennen. Als Extensität erkennen wir alsbald die Gewichtsmenge […] Und als Intensität erkennen wir die Höhe, zu welcher das Gewicht gehoben wird” (ibid. p. 117).

[74] “*Materielle Faktoren* nenne ich die fraglichen Größen deshalb, weil durch sie der alte Begriff der Materie bedingt wird” (ibid. p. 111).

[75] “Da ein fester Körper seine Gestalt ändern kann, ohne sein Volum zu ändern, so liegt hier eine unterschiedliche, wenn auch ähnliche Art Energie vor, welche wir […] *Formenergie* nennen wollen. In der Physik ist diese Eigenschaft als Elastizität bekannt” (ibid. p. 114).

[76] “Hier haben wir die erste Energieart, die unserem Körper zukommt. Wir nennen sie *Volumenergie*, weil sie sich mit dem Volum des Körpers ändert” (ibid. p. 113).

[77] “Den Raum haben wir als das Betätigungsgebiet der Energien auffassen gelernt; es kann uns daher nicht wundernehmen, daß neben der Volum- und Formenergie, die den dreidimensionalen Raum erfüllt, auch noch eine Oberflächenenergie […] und eine lineare oder *Distanzenergie* besteht, welche sich in den Gravitationswirkungen geltend macht” (ibid. p. 118).

[78] “So sehen wir die Materie überflüssig werden, weil wir sie analysiert und ihre Bestandteile erkannt haben” (ibid. p. 124).

[79] “Den Wärmeinhalt der Körper pflegen wir nicht zur Materie zu rechnen, obwohl es sich ebenso um eine besondere Energieart handelt […] Es liegt dies daran, daß wir den Extensitätswert der Wärme, die Entropie, so gut wie gar nicht kennen” (ibid. p. 124). “Die *chemische* Energie […] gehört ebenso wie die vorher genannten Grundenergien zu dem eisernen Bestande eines jeden Körpers und hat ausgesprochen “materiellen” Charakter. Man muß es nur der Unkenntnis der chemischen Erscheinungen zu der Zeit, wo der Begriff der Materie festgestellt wurde, zuschreiben” (ibid. p. 125).

energy subsumes matter and properties of matter and is, consequently, more general than the concept of matter.

Let us move on to the living being issue, where the main topic is the human being. From an energetic point of view, the living being is in a permanent exchange of energy with the outside.[80] Taken as an energy system, chemical energy is the most fundamental energy of a living being. The food it uses is made up of chemical energy. Our muscles work thanks to chemical energy and even the mode of action of nerves is linked to this form of energy. Most likely, Ostwald adds, the very phenomena of memory have their origin in chemical relations.[81] The author notes that he is approaching the issues with the means available to science,[82] which in some cases were scarce, because the research was very recent.[83]

His approach to psychic phenomena in general develops under two headings: on the one hand the traditional discussion between matter and spirit; on the other hand, the relation of spirit to the concept of energy. According to Ostwald, what is traditionally related with the spirit can be expressed in energetic terms[84]: the impressions of the senses are seen as passage of energy; reception by the nerves is understood as transformation of energy; communication through the nerves is interpreted as propagation of energy.[85] The energy of the nerves is said to be psychic energy, whose basis is chemical energy.[86] Since psychic phenomena can be given

---

[80] "Von unserem Standpunkte ist ein wesentliches, wenn auch nicht das zureichende Kennzeichen des Lebens die *beständige Energiebetätigung*" (ibid. p. 129).

[81] "Unsere Nahrungsmittel bestehen aus chemischer Energie [...] Unsere Muskeln arbeiten mit chemischer Energie und ebenso ist die noch so geheimnisvolle Wirkungsweise der Nerven gleichfalls mit dieser Energieart auf das engste verbunden. Vor allen Dingen aber beruht aller Wahrscheinlichkeit nach eine besondere Eigentümlichkeit aller Lebenserscheinungen gleichfalls auf chemischen Verhältnissen, nämlich die Erscheinung des *Gedächtnisses* im allgemeinsten Sinne, wie sie zuerst von *E. Hering* erkannt worden ist" (ibid. p. 134).

[82] "die Natur verfügt selbstverständlich über den gesamten Bestand an Möglichkeiten [...] Unsere Erklärungsversuche müssen wir dagegen mit dem augenblicklich bekannten Bestand unserer Wissenschaft Machen" (ibid. p. 136).

[83] "Ich bin gern bereit, anzuerkennen, daß die Ähnlichkeiten zunächst nur sehr oberflächlicher Natur sind; doch muß andererseits erwogen werden, daß das Gebiet chemischer Erscheinungen, das hier in Betracht kommt, erst seit einem Jahrzehnt der systematischen Bearbeitung unterworfen worden ist, und daß daher nur kleine Teile desselben inzwischen bekannt geworden sind" (ibid. p. 135).

[84] "[...] glaube ich so auffassen zu dürfen, *daß die geistigen Geschehnisse ebenso sich als energetische auffassen und deuten lassen, wie alle übrigen Geschehnisse auch*" (ibid. p. 144).

[85] "Nun haben wir bereits gesehen, daß ein Sinneseindruck ganz allgemein beschrieben werden kann als ein Energieübergang zwischen der Außenwelt und einem Körperteil, der durch besondere Organisation empfindlich für kleine Energieunterschiede gemacht worden ist. Die Tatsache, daß verschiedenartige Energien, die auf den gleichen Apparat wirken, doch Empfindungen gleicher Art auslösen (z.B. Lichterscheinungen durch mechanische Einwirkung auf den Sehnerven), erfordert die Deutung, daß bereits im Sinnesapparat eine Umformung der äußeren Energie in eine andere Form stattfindet, welche durch den Nerv fortgepflanzt wird [...] Wir wissen aber, daß irgendeine *Energie* fortgepflanzt wird [...] Wir wollen also der Kürze wegen von *Nervenenergie* reden" (ibid. p. 145).

[86] "Wohl aber wissen wir, daß die Quelle dieser psychischen Energie *chemischer* Natur ist" (ibid. p. 153).

in terms of energy and the author had already shown that the concept of matter was dispensable, the matter-spirit dilemma would no longer arise.[87] Energetics is able to account for phenomena that were traditionally attributed to one or the other.[88]

The last chapter of the book is entitled "the sociological energetics". The subject is man in society. Energy emerges in complex organizational activities, but also in simple situations, as, for example, using a stick to reach something. Ostwald explains that through the stick's form of energy, muscular energy is transformed and communicated there, where the stick reaches.[89] Cutting objects, a knife or a sword, are other examples of transformers of energy. Here muscle energy is concentrated on a thin surface, increasing the intensity of the pressure.[90] In accordance with the purposes and cultural development of societies, the means can be very complex: animals, men, machines, fuels, etc.[91] Ostwald puts as a general task to the whole culture, to find the transformation coefficients of the most favorable energies.[92]

This task appears with an ethical dimension in his book *The Energetic Imperative* (Ostwald, 1912b). This imperative simply states, *don't waste energy, harness it.* This wording alone could lead one to think that it is a warning not to use energy unnecessarily. However, the imperative also concerns another part of human action. If, for example, in a teacher's competition the jury chooses the candidate least able for the job, the jury is wasting energy in the sense of Ostwald.

In the first of the two parts of the 1908 book, a history of energetics is presented. Mayer was the first energeticist. The history of energetics began in 1842, with Mayer's first writing. Nevertheless, there was a prehistory, so to speak, that takes Ostwald to the Greeks.[93] Relevant points in the history of energy before 1842 were: Jean Bernoulli's proposition about the balance of forces, where energy was understood as

---

[87] "Es besteht [...] gar nicht mehr die Aufgabe, zu ermitteln, wie Geist und Materie in Wechselwirkung treten können, sondern es entsteht die Frage, *wie sich der Begriff der Energie, der viel weiter als der der Materie ist, zu dem Begriff des Geistes stellt* " (ibid. p. 144).

[88] *"für die mechanistische Weltauffassung besteht zwischen den physischen Erscheinungen als mechanischen einerseits und den geistigen andererseits eine unüberbrückbare Kluft; für die energetische Weltauffassung besteht im Gegenteile ein stetiger Zusammenhang zwischen den einfachsten Energiebetätigungen, den mechanischen, und den verwickeltsten, den psychischen"* (ibid. p. 156).

[89] "Nehmen wir das einfachste aller Werkzeuge, den abgebrochenen Baumast [...] Durch die Formenergie des Stabes [...] wird die Muskelenergie transformiert und dorthin übertragen, wo der Stab auftritt" (ibid. p. 161).

[90] "Eine andere Art von Transformatoren finden wir in den *schneidenden* Werkzeugen vor. Hier wird durch die schmale Druckfläche der Schneide eine Konzentration der Muskelenergie auf diese lineare Berührungsfläche und eine entsprechende Steigerung der Intensität des Druckes bewirkt" (ibid. p. 161).

[91] "Die nächste Stufe ist die Aneignung fremder Arbeit für eigene Zwecke [...] ist die Verwendung anderer *Menschen* in solchem Sinne [...] Die dritte Stufe in der Besitzergreifung fremder Energien ist endlich die der *anorganischen*" (ibid. p. 162).

[92] "Man darf wirklich *als die allgemeine Aufgabe der gesamten Kultur die hinstellen, die Transformationskoeffizienten der umzuwandelnden Energien so günstig wie möglich zu gestalten*" (ibid. p. 165).

[93] "So kann man zwar angeben, daß die klare Feststellung allgemeinen Begriffes der Energie nicht früher als im Jahre 1842 erfolgt ist, und könnte somit, wenn man genaue geschichtliche Kapiteleinteilungen liebt, die Geschichte der Energielehre oder Energetik von diesem Zeitpunkte ab beginnen.

the product of force by the measured path[94]; the recognition of the impossibility of perpetual motion; and Mechanics that subsumed statics and dynamics through forms of energy, work and energy of motion, respectively.[95]

The fourth chapter of the book is on "the mechanical equivalent of heat". Mayer, whose article (1842) is transcribed (Ostwald, 1912a, pp. 52–8), considered mechanical work and heat as forms of one and the same thing.[96] Ostwald interprets this 'same thing' as a kind of substance, which, he adds, had been fundamental for his own research.[97] Joule's explanation of heat, consisting of a motion of particles of matter, atoms or molecules, is considered an ad hoc hypothesis.[98] Helmholtz addressed the development of heat in animals as Mayer did, but used, however, Joule's mechanistic hypothesis as his fundamental idea.[99] (Ostwald's objection to this hypothesis stems from the difficulty in explaining psychic phenomena mechanically.[100]) Rankine gave

---

Aber ebensowohl muß man zugeben, daß die ersten Ansätze zur Gestaltung dieses Begriffes [...] bis zu den griechischen Mathematikern und Naturphilosophen zurückverfolgen lassen" (ibid. p. 9).

[94] "Doch der klare Ausspruch des Prinzips, der sogar bezüglich der Bezeichnungsweise ganz modern ist, findet sich erst in einem Briefe, den *Jean Bernoulli* im Jahre 1717 an *Varignon* geschrieben hat. Die entscheidende Stelle lautet: "En tout équilibre de forces quelconques, en quelque manière qu'elles soient appliquées, et suivant quelques directions qu'elles agissent les uns sur les outres, ou médiatement, ou immédiatement, la somme des énergies affirmatives sera égale à la somme des énergies négatives, prises affirmativement." [...] Als Energie wird ausdrücklich das Produkt der Kraft in dem durchmessenen Weg, letzterer in der Richtung der Kraft gerechnet, definiert" (ibid. p. 16–7).

[95] "So definieren wir besser die Statik als die *Wissenschaft von der Energieform, welche man Arbeit nennt*, und haben sachgemäß zu fragen, ob für die Dynamik vielleicht eine ähnliche Definition vorhanden ist. Die Antwort lautet bejahend. Auch für dieses zweite Gebiet der Mechanik wird sich als Zentralbegriff eine bestimmte Energieart herausstellen, welche wir die *Bewegungsenergie* nennen werden" (ibid. p. 29).

[96] "*Man müßte sich nur entschließen*, zwei *so grundverschiedene Dinge, wie mechanische Arbeit und Wärme als zwei Formen desselben Wesens zu betrachten.* Zwar ist die Physik, wie sie auf der Universität gelehrt worden war, himmelweit von einem solchen Gedanken entfernt" (ibid. p. 48).

[97] "Für unsere allgemeine Untersuchung ist das Wesentlichste, was *Mayer* geleistet hat, die *substanzielle* Auffassung dessen, was er Kraft nennt, d. h. der Energie" (ibid. p. 58).

[98] "Er [Joule] betrachtete das konstante Verhältnis zwischen Arbeit und Wärme, welches er durch seine Versuche bestätigt hatte, als ein Zeichen dafür, daß auch die Wärme im Grunde mechanischer Natur sei und daß sie, entsprechend einer bereits damals weit verbreiteten Anschauung, in einer Bewegung der kleinsten Teilchen der Materie, der Atome oder Moleküle, bestehe. Alsdann würde es sich bei der Umwandlung von Arbeit in Wärme um einen ganz ähnlichen Vorgang handeln, wie bei der Umwandlung von Arbeit in lebendige Kraft gewöhnlicher oder sichtbarer Art, und damit wäre das Äquivalentsgesetz "erklärt". Allerdings ist diese Erklärung von derselben unbefriedigenden Beschaffenheit, wie alle ad hoc erfundenen Hypothesen" (ibid. p. 62–3).

[99] "Der dritte Forscher, welchem wir Wesentliches für die Ein- und Durchführung des Gesetzes von der Erhaltung der Energie verdanken, ist *H. Helmholtz*. Er ist auf gleichem Wege wie *Mayer*, nämlich durch Nachdenken über das Problem der Wärmeentwicklung im tierischen Körper, zu seinen Schlüssen gelangt [...] Als Grundgedanken benutzt er den gleichen, wie *Joule*, nur daß er ihn mathematisch zu vertiefen weiß" (ibid. p. 64).

[100] "Daß die Natur auf die Mechanik bewegter Atome zurückzuführen sei, galt nicht als eine noch des Beweises bedürftige Hypothese, sondern als ein Postulat der wissenschaftlichen Forschung [...] Erst

an impetus to energetics, but he was also committed to the mechanistic hypothesis.[101] Hence, Ostwald does not consider him an energeticist. An imprint of this retrograde perspective had remained in the distinction between actual and potential energy.[102] Nevertheless, he recognizes Rankine's merit of having observed that the various energies are given as products of two factors.[103] (This remark by Rankine was crucial for Ostwald's approach to matter and energy, which led to the elimination of the concept of matter.)

Conclusion

Ostwald used Mayer's concept of energy (force), substantialized it, and generalized it to all scientific domains and to ethics. For this generalization, he uses the forms of energy (chemical, psychic, etc.) and the idea of an 'energy transformer'. It is through this concept that he generalizes energy to ethics. If in a competition for teachers, the example given above, the candidate with the least capacity for the job is chosen, the worst energy transformer, among those available, was chosen.

There is a reason for Ostwald's interest in Energetics. He advocates this theory because the concept of energy is fundamental in scientific research. In a controversy with Boltzmann (1896a, b), he further points out that he defends energetics because the concept of energy had been crucial to his own research (Ostwald, 1896).[104]

References

Boltzmann, L. (1896a). Ein Wort der Mathematik an die Energetik. *Annalen Der Physik, 57*, 39–71.
Boltzmann, L. (1896b). Zur Energetik. *Annalen Der Physik, 58*, 595–598.
Helm, G. (1896). Zur Energetik. *Annalen Der Physik, 57*, 646–659.

---

in neuester Zeit erwies es sich als nötig, diese Prüfung vorzunehmen, namentlich nachdem der mechanische Materialismus bezüglich der psychischen Erscheinungen zu auffallenden Widersprüchen geführt hatte" (ibid. p. 66).

[101] "daß der erste, welcher das Wort gebildet hat und auch einen gedanklichen Vorstoß in solchem Sinne versucht hat, doch nicht eigentlich selbst als Energetiker bezeichnet werden darf. Dieser Mann ist [...] *Rankine* [...] Was ihn aber wesentlich zu seinem Nachteile von *Mayer* unterscheidet, ist daß er ganz und gar bei der mechanistischen Hypothese von *Joule* und *Helmholtz* stehen bleibt und dadurch die wesentlichste Seite der eigentlichen Energetik, die Freiheit von Hypothesen, verfehlt" (ibid. p. 98-9).

[102] "Ein Nachbleibsel dieses ungenügenden Standpunktes ist bis auf den heutigen Tag in der wissenschaftlichen Nomenklatur des Gebietes erhalten geblieben, nämlich in der unzweckmäßigen Unterscheidung von *aktueller* und *potentieller Energie*, welche *Rankine* eingeführt hatte, und welche alsdann von *Thomson* und vielen anderen Schriftstellern des Gebietes wieder aufgenommen worden ist" (ibid. p. 99).

[103] "Kann insofern den Gedanken *Rankines*, die übrigens durchweg der Klarheit ermangeln, kein Wert für die Entwicklung der Energetik zugeschrieben werden, so erfordert doch die Gerechtigkeit [...] Es ist dies die Beobachtung, daß die verschiedensten Energien übereinstimmend *sich als Produkt zweier Faktoren darstellen lassen*" (ibid. p. 100-1).

[104] See also Planck (1896) and Helm (1896); and Hiebert (1971).

Hertz, H. (1894). *Die Prinzipien der Mechanik in neuem Zusammenhange dargestellt* (P. Lenard, Ed. Vol. III). J. A. Barth.
Hiebert, E. N. (1971). The energetics controversy and the new thermodynamics. In D. H. D. Roller (Ed.), *Perspectives in the history of science and technology* (pp. 67–86). University of Oklahoma Press.
Lodge, O. (1879). An attempt at a systematic classification of the various forms of energy. *Philosophical Magazine, 8*, 277–286.
Lodge, O. (1885). On the identity of energy: In connection with Mr Poynting's paper on the transfer of energy in an electromagnetic field; and the two fundamental forms of energy. *Philosophical Magazine, 19*, 482–494.
Maxwell, J. (1873). *A treatise of electricity and magnetism*. Clarendon Press.
Ostwald, W. (1896). Zur Energetik. *Annalen Der Physik, 58*, 154–165.
Ostwald, W. (1912a). *Die Energie* (2 ed.). J. A. Barth.
Ostwald, W. (1912b). *Energetische Imperative*. J. A. Barth.
Planck, M. (1896). Gegen die neuere Energetik. *Annalen Der Physik, 57*, 72–78.
Planck, M. (1921). *Das Prinzip der Erhaltung der Energie* (4 ed.). Teubner.
Poincaré, H. (1897). Les idées de Hertz sur la mécanique. *Revue générale des Sciences, 8*, 734–743.
Poynting, J. H. (1884). On the transfer of energy in the electromagnetic field. *Philosophical Transactions of the Royal Society of London, 175*, 343–361.
Tait, P. (1885). *Properties of matter*. Adam & Charles Black.
Thomson, W., & Tait, P. (1867). *Treatise on Natural Philosophy, Vol. 1*. Oxford University Press.

Chapter 5
Trends in Contemporary Textbooks

'Trends in contemporary textbooks' might lead one to think that the focus of this chapter is on physics teaching. Contemporary textbooks appear here, however, because they present definitions of the concept of energy and formulations of the conservation principle. It is these definitions and formulations that will be analyzed in the first two sections of this chapter. (The principles will be addressed first, because the most used concept of energy comes from the conservation principle.) The third section deals with the references to Mayer and Joule used by textbook writers when introducing the issue of energy. Throughout the chapter, connections between the propositions addressed and the corresponding ones from the nineteenth century are made.

5.1 Principles

This section is titled 'principles' and not 'principles of conservation of energy' because not all principles that follow, which concern the heat-work relationship, use the concept 'energy'.

5.1.1 Principle of Equivalence

At the beginning of the twentieth century, 'principle of equivalence' was a common concept for the first law of thermodynamics. In 1919, Preston writes:

> "The modern science of thermodynamics is based on two fundamental principles [...] The first of these is the principle of equivalence". (Preston, 1919, 667)[1]

[1] Voigt used the same idea: "*Die so verallgemeinerte Energiegleichung kann als die exakte Formulierung des Satzes von der Äquivalenz von Wärme und Arbeit gelten*" (Voigt, 1903, p. 78).

R. Lopes Coelho, *What Is Energy?*, History of Physics,
https://doi.org/10.1007/978-3-031-51855-3_5

Müller and Pouillet (1926), preferred the concept 'principle of equivalence' to 'principle of conservation of energy'. For them, the latter was valid because the former was valid. Furthermore, they wrote:

> "Energy is regarded as indestructible. The principle of energy is therefore at first not an empirical law, but a *postulate*, which is, however, quite consistent with the facts of experience (law of equivalence)"[2]

'Principle of energy' in the quote refers to 'principle of conservation of energy', since energy is taken as indestructible, therefore, constant in quantity. This principle as such does not come from experiments, according to these authors. Only the equivalence is based on experiments. The principle of equivalence has another advantage, according to these authors. Stating that heat and work are equivalent, it avoids assumptions about the nature of heat.[3] Due to the advantages of the equivalence principle, one would ask why consider the first law of thermodynamics in terms of energy. There is only an epistemological reason for that:

> "Since the main aim of physical science is to obtain a uniform, simple overall picture, one is forced to go beyond the formal, purely descriptive principle of equivalence, even if in this step one has to leave the ground of immediate experience".[4]

The concept 'principle of equivalence' has fallen into disuse, although Allen and Maxwell still used it in 1962:

> "J. P. Joule was the first to establish on a satisfactory experimental basis the Principle of Equivalence. This principle may be expressed by saying that when exchange occurs between work and heat, the *ratio* of the exchange is fixed". (Allen & Maxwell, 1962, p. 284)

5.1.1.1 Historical Connection

Clausius' and Thomson's first law of the theory of heat established the validity of the mechanical equivalent of heat (Sect. 3.1). In the 1850s, a search begins to determine the mechanical equivalent of heat through new methods (Helm, 1898, p. 34; Kipnis, 2014, p. 2028). The convergence of the results obtained led to the idea that the equivalent was valid in general. This validity was expressed by 'principle of equivalence'.

The French physicist Émile Verdet taught in his courses: "The principle of equivalence of heat and work is the first fundamental principle of the mechanical theory

---

[2] "Energie wird als unzerstörbar angesehen. Das Energieprinzip ist somit zunächst kein empirisches Gesetz, sondern ein *Postulat*, das sich allerdings mit den Erfahrungstatsachen (Äquivalenzgesetz) durchaus im Einklang befindet". (Müller & Pouillet, 1926, p. 126)

[3] Die "Formulierung des ersten Hauptsatzes, die von jeder Hypothese, etwa über die Natur der Wärme frei ist, besagt daher einfach: Wärme und mechanische Arbeit sind äquivalent" (ibid. p. 109).

[4] "Da nun gerade der Gewinn eines einheitlichen, einfachen Gesamtbildes das Hauptziel der physikalischen Wissenschaft bildet, ist man gezwungen, über das formale, rein beschreibende Äquivalenzprinzip hinauszugehen, wenn auch bei diesem Schritte der Boden der unmittelbaren Erfahrung verlassen werden muß". (ibid. p. 126)

of heart".[5] The demonstration of the principle he gave is based on experimental determinations of the mechanical equivalent of heat (Joule, Hirn and Favre) and the convergence of the values obtained (Verdet, 1868, pp. 39–74).

In the *Thermodynamics* course (1888–89), Poincaré has a chapter on the equivalence principle. In the demonstration of the principle, he uses a set of experimental results on the mechanical equivalent of heat (Joule, Hirn and Violle) and shows that the values obtained conform to each other. It is this invariability of the equivalent, says Poincaré, that constitutes the demonstration of the principle.[6] He continues, this principle has the advantage over the principle of conservation of energy of not making any assumptions about the molecular constitution of bodies.[7]

5.1.2 Perpetuum Mobile

Some authors explained the principle of conservation of energy through the *perpetuum mobile* or even took the impossibility of *perpetuum mobile* as an expression of the first law of thermodynamics.

In the *Treatise on Heat*, Saha and Srisvastava present the principle of conservation of energy as the result of human experience, the impossibility of a perpetual motion machine.

> "It follows from a result of human experience which may be stated in this form: "It is impossible to design a machine which will create energy out of nothing and produce perpetual motion. Energy can only be transformed from one form to the other."" (Saha & Srisvastava, 1935, p. 434)

The authors do not tell us how the impossibility of perpetual motion led to the principle of conservation of energy. Indeed, they use the idea of the impossibility of creating motion out of nothing and replace 'motion' with 'energy'.

In Wolf's *Elements of Physics*, the impossibility of a *perpetuum mobile* is a way of expressing the first law of thermodynamics: "the first law can be expressed in the short form: *A perpetuum mobile is not possible*".[8]

---

[5] "Le principe de l'équivalence de la chaleur et du travail est le premier principe fondamental de la théorie mécanique de la chaleur". (Verdet, 1868, Vol. 7, p. 39)

[6] "Cette invariabilité de E constitue précisément le *principe de l'équivalence*". (Poincaré 1892, p. 58)

[7] "Cette manière d'envisager ce principe, aujourd'hui généralement adoptée, présente l'avantage de ne faire aucune hypothèse sur la constitution moléculaire des corps". (Poincaré 1892, p. 65) Poincaré arrives at the principle of conservation of energy from the hypothesis of central forces. Hence the reference to the hypothesis of molecular constitution in the quote.

[8] "Man hat für derartige Phantasiemaschinen den Namen geprägt. Daher wird schließlich der erste Hauptsatz auch gern in der kurzen Form ausgesprochen: *Ein Perpetuum mobile ist nicht möglich*". (Wolf 1949, p. 273)

5.1.2.1 Historical Connection

In 1775, the Academy of Sciences in Paris passed the resolution not to examine "any machine announced as showing perpetual motion" (Elkana, 1974, p. 30) Conservation of energy does not come, however, from this experience with machines. The only one of the four so-called discoverers (Chap. 2) who mentions the impossibility of perpetual motion is Colding. However, he does not use the impossibility of *perpetuum mobile* to justify his thesis, but rather the opposite: "this thesis seems so pressingly necessary for a complete proof of the impossibility of a Perpetuum Mobile that without it every such proof must be regarded as false" (Colding, 1856, p. 14).[9]

Expressing the first law of thermodynamics through the impossibility of a *perpetuum mobile* leaves out an important aspect of the law, which is the mechanical equivalent of heat, for which, in fact, the law was created.

5.1.3 Principle of Conservation of Energy

The most common formulation of the principle of conservation of energy uses three concepts that express properties of energy: uncreatable, indestructible and transformable.

Hudson and Nelson (1982) write:

> "energy can only be changed from one form to another; it cannot be created or destroyed. This conclusion, based on experiment, is known as the law of conservation of energy". (Hudson & Nelson, 1982, p. 95)

If energy cannot be destroyed, then one could infer that energy is something real. The authors, however, point out that energy is not a substance: "Although it appears in many different forms, it is not a physical substance, but a calculated quantity" (ibid. p. 95).

Breithaupt stresses that energy exists in different forms, which may lead one to think that it is something real:

> "Energy exists in different forms"
>
> "Energy can be transformed from any one form into other forms. Whenever energy changes from one form into other forms, the total amount of energy is unchanged. This general rule is known as the principle of conservation of energy". (Breithaupt, 1999, p. 157)

Hecht used 'transformation' and 'transfer' in the formulation of the principle:

> "Energy can neither be created nor destroyed, but only transferred from one system to another and transformed from one form to another". (Hecht, 2000, p. 571)

---

[9] Helmholtz assumed that it is impossible to get a lasting motive force out of nothing. Caneva addresses the relation of this principle to the impossibility of *perpetuum mobile*. (Caneva, 2021, pp. 1–4). Helmholtz, however, refers his principle to Carnot and Clapeyron (Sect. 2.4).

This formulation could lead to the idea of energy as an entity. The author, however, points out that energy "does not have a separate existence. It does not spontaneously vanish at one place in the Universe and simultaneously appear at another" (ibid. p. 571).

Dransfeld, Kienle and Kalvius write:

> "Energy can be neither produced nor destroyed but remains constant in every closed system. (Fact of experience!)"[10]

This proposition concerns only the conservation of energy (no change in quantity). Further on, they include 'transformation':

> "Energy occurs in various forms, such as gravitational energy, kinetic energy, thermal energy, elastic energy, electrical energy, magnetic energy, chemical energy, nuclear energy, and radiant energy. In the processes taking place in nature, one form of energy is always transformed into another, but the sum of the component energies remains constant"[11]

In *Sears and Zemansky's Physics*, Young and Freedman present the law of conservation of energy as follows:

> "energy is never created or destroyed; it only changes form". (Young & Freedman, 2004, p. 264)

Çengel and Ghajar introduce the principle in this same way:

> "Energy can neither be created nor destroyed during a process; it can only change forms". (Çengel & Ghajar, 2015, p. 11)

5.1.3.1 Historical Connection

The most common formulation of the principle of conservation of energy originates in Mayer's 1842 paper (Sect. 2.1.1.1). Two points, however, are important to note. First, what was taken from Mayer are the properties of force—indestructibility and transformability—and not the principle he formulated: "In all physical and chemical processes, *the given force remains a constant magnitude*" (Mayer, 1845, p. 32). Second, today's formulation can mislead us to think of energy as substantial, whereas Mayer pointed out that force (energy) is not material. According to him, indestructibility concerns only the quantity of force and transformability was not designed to convey knowledge about the phenomena beyond what is observable (Sect. 2.1.2).

---

[10] "*Energie* kann weder erzeugt noch vernichtet werden, sondern *bleibt in jedem abgeschlossenen System konstant*. (Erfahrungstatsache!)". (Dransfeld et al., p. 109)

[11] "Energie kommt in verschiedenen Formen vor, wie z.B. als Gravitationsenergie, kinetische Energie, Wärmeenergie, elastische, elektrische, magnetische, chemische Energie, Kernenergie und Strahlungsenergie. Bei den in der Natur ablaufenden Prozessen wird immer eine Energieform in eine andere umgewandelt, wobei jedoch die Summe der Einzelenergien konstant bleibt". (ibid. p. 109)

5.1.4 First Law of Thermodynamics

Tipler and Mosca formulate the first law of thermodynamics, which they consider as "a statement of the conservation of energy", as follows:

> "The change in the internal energy of the system equals the heat transferred into the system plus the work done on the system". (Tipler & Mosca, 2004, p. 566)[12]

The law is coupled with a remark regarding the use of the terms work and heat:

> "It is correct to say that a system has a large amount of internal energy, but it is not correct to say that a system has a large amount of heat or a large amount of work". (ibid. p. 567)

Some authors explain what internal energy consists of:

> "Internal energy U is the energy associated with the microscopic components of a system – the atoms and molecules of the system". (Faughn et al., 2006, p. 253)
>
> "The internal energy of a substance is the sum of the molecular kinetic energy […], the molecular potential energy […] and other kinds of molecular energy". (Cutnell & Johnson, 2006, p. 243)

Nevertheless, according to Chabay, Sherwood and Titus,

> "internal energy is a convenient catch-all term that denotes "forms of energy that we choose not to analyze in detail at the moment," and not a fundamental form of energy". (Chabay et al., 2019, p. 507)

Hecht points out that "Internal energy cannot be observed; it cannot be measured as it presumably exists." (Hecht, 2019, p. 500).

5.1.4.1 Historical Connection

In reorganizing the theory of heat, Clausius assumed the validity of the mechanical equivalence of heat. This becomes its first law of thermodynamics. In the contemporary formulation of this law, the mechanical equivalent is only presupposed. The content of the law is given by means of the internal energy, which refers to Thomson's mechanical energy of a body.[13]

Thomson was interested in knowing the total amount of energy in a body. Since this was not possible, he defined the mechanical energy of a body in relation to a standard state (Sect. 3.2). Maxwell called it 'intrinsic energy' (Sect. 3.6). This adjective clearly refers to the interior of the body. Contemporary authors keep the idea of the interior of the body, with the term *internal* energy. Whereas Thomson assumed that heat exists in a body, contemporary authors keep the word 'heat' for the relationship with the surroundings:

---

[12] Similar formulations of the first law can be found in other modern physicists (Kane & Sternheim, 1988, p. 262; Halliday et al., 1997, p. 467; Wilson & Buffa, 1997, p. 386).

[13] 'Mechanical energy' refers nowadays to the sum of kinetic and potential energy in Mechanics. Some physicists also use the concept 'principle of conservation of mechanical energy' (Halliday et al., 1997, p. 160–161; Buffa & Wilson, 1997, p. 150; Serway & Beichner, 2000, p. 215).

> "*it is not correct to say that a substance contains heat*. The substance has internal energy, not heat. The word "heat" is used only when referring to the energy actually in transit from hot to cold". (Cutnell & Johnson, 2006, p. 243)

The idea of energy in transit has nothing to do with Thomson.[14] It refers to Lodge's claim that energy is transferred between bodies or between bodies and the aether.

The contemporary equation of the law is

$$\Delta E_{int} = Q_{in} + W_{on} \tag{5.1}$$

where E_{int} refers to internal energy, Q to heat and W to work. (Tipler & Mosca, 2004, p. 566) This goes back to Thomson and Clausius. The variation in Thomson's mechanical energy is given as

$$de = JQ - W \tag{5.2}$$

where J stands for the mechanical equivalent of heat (Sect. 3.2). Clausius said that this energy corresponds to his function U.[15] This function is given as $U =$ *the heat existing in the body* + *the internal work*. (Clausius, 1879, p. 27, eq. II; p. 31, Eq. 3) It appears in his equation (ibid. p. 31, eq. III)

$$dQ = dU + dW \tag{5.3}$$

where dQ stands for an indefinitely small quantity of heat imparted to the body and W for work.

5.2 Concepts of Energy

Some textbooks present the principle of conservation but do not an explicit definition of energy. Others define energy; and still others claim that we do not know what energy is. The concepts will be addressed in this order.

5.2.1 Indestructible and Transformable

The principle that states that energy cannot be created or destroyed but only transformed provides us with the properties of energy. Energy is that something which

---

[14] Thomson could not agree with this meaning for the concept he had introduced (see Smith, 1998, p. 289).

[15] "The function *U*, first introduced by the author in the above-mentioned paper [Clausius 1850], has been since adopted by other writers on Heat [...] Thomson, in his paper of 1851 called it *the mechanical energy of a body in a given state*". (Clausius 1879, p. 31)

is uncreatable, indestructible and transformable. Some physicists understand energy as a real entity with these properties (Sect. 5.1.3). Others have been careful to point out, however, that energy is not an entity (Feynman et al., 1963; Hudson & Nelson, 1982; Hecht, 2000, among others).

The authors who assume that energy is transformable, use the concept 'forms of energy' in the interpretation of phenomena. According to them, heat is a form of energy.

Heat as a Form of Energy

'Transformation' means change of form. Therefore, what appears at the beginning of a transformation of energy and at the end must be forms of energy. If motion gives rise to heat, for example, motion and heat are consequently forms of energy. There is, therefore, a conceptual reason to take heat as a form of energy.

Nolting (2002) presents the idea concisely: "*Heat*" = "*form of energy*" (Nolting, 2002 p. 148).[16] The same idea appears in other textbooks: Hund (1956),[17] Aktins (1986),[18] Hänsel and Neumann (1993),[19] Benenson et al. (2001)[20] and Böge and Eichler (2002).[21]

Cassiday, Holton and Rutherford show the link between 'transformation' and 'form' of energy:

> "The disappearence of kinetic energy [...] is accompanied by the appearance of *heat*. This suggests, though by no means proves, that the kinetic energy [...] was transformed into heat. If so, heat must be one form of energy". (Cassiday et al., 2002, p. 253)

Criticism

Some physicists criticize the concept of heat as a form of energy or simply adopt another solution. According to Allen and Maxwell (1962), who still used the principle of equivalence, heat as a form of energy was just a hypothesis.

> "When such an exact relation between mechanical energy and heat has been established by experiment, it is not difficult to take a further step by introducing the hypothesis that heat is itself a form of energy, so that in the exchanges considered there is a transmutation from one form of energy to another". (Allen & Maxwell, 1962, p. 285)

---

[16] "Ein wesentlicher Bestandteil des Ersten Hauptsatzes ist deshalb die Aussage: "*Wärme*" = *Energieform*". (Nolting, 2002, p. 148)

[17] "Die genauere Untersuchung dieser Erscheinungen führte zu der Ansicht, daß Wärme eine Energieform sei. Zunächst führte sie zu der Ansicht, Wärme sei Bewegung der kleinsten Teile der Körper und damit eine Form von Bewegungsenergie (*F. Mohr* 1837); später zu der Ansicht, Wärme sei eine Form der Energie [...] (*R. Mayer* 1842)". (Hund, 1956, pp. 49–50)

[18] "Wärme ist eine Form von Energie". (Aktins, 1986, p. 233)

[19] "*Robert Mayer* erkannte als erster (1842), daß die Wärmemenge eine Form der Energie ist". (Hänsel & Neumann, 1993, p. 222)

[20] "Heat is a particular form of energy connected with the temperature increase of a substance". (Benenson et al., 2001, p. 678).

[21] "Wärme ist eine Energieform". (Böge & Eichler, 2002, p. 83)

In Bergmann and Schaefer's *Experimental Physics* (1998), the term 'forms of energy' is taken as somewhat *misleading*. They explain: the different 'manifestations' or 'types' of energy originate from different ways in which it is transported (work, heat, radiation) or it is associated with matter (kinetic, potential, chemical, electrical, etc. energy).[22]

5.2.1.1 Historical Connection

The divergence between physicists about the question of whether energy is a substance or not appeared in the nineteenth century. Planck had expected that the concept of energy as a substance would one day be overcome and Hertz argued that this concept is logically unacceptable. Nevertheless, Lodge and Ostwald claimed that energy is something real.

The concept 'forms of energy' comes from Mayer's forms of force. In 1845, he presents 5 forms of force and points out the 25 transformations with these forms: 20 are transformations between different forms, as motion into heat, and 5 are transformations of the same form like motion into motion (Sect. 2.1.2).

5.2.2 Energy Transfer

Preston tells us that energy moves through space and the aether was postulated to serve as a vehicle for energy.

> "Energy is again often stated to be only associated with matter, so that matter has been defined as the vehicle of energy; this, however, does not hold according to our limitation of the word matter, for we know that energy in immense quantities is perpetually passing through space with the velocity of light [...] The ether, then, is the great vehicle of energy; and, indeed, it is chiefly on this account that the ether has been postulated". (Preston, 1919, p. 80)

Phenomena such as throwing a body in the air or the motion of a pendulum are seen as energy exchanges between bodies and the ether

> "An evident reply to the question of what becomes of the motion of a projectile rising upwards is that it passes into the ether. [...] The oscillation from kinetic to potential, and from potential to kinetic, in the case of the pendulum is then, from this point of view, merely an interchange of energy of motion going on between the mass of the pendulum and the ether around it. [...] The oscillation of energy, then, is from ether to matter, and from matter to ether, and on this oscillation all the physical life of the universe depends". (ibid. p. 88)

---

[22] "Es gibt im Grunde nur eine wohldefinierte Größe Energie, die in einem Körper oder in einem Raumbereich vorhanden ist. Verschiedene "Erscheinungsformen" oder "Arten" dieser Energie beruhen entweder auf der unterschiedlichen Art und Weise, wie sie transportiert wird (Arbeit, Wärme, Strahlung) oder auf den verschiedenen Möglichkeiten, wie sie mit Materie verbunden sein kann (kinetische, potentielle, chemische, elektrische usw. Energie). Es ist daher etwas irreführend, von verschiedenen *Energieformen* zu sprechen, wie es oft geschieht. Denn es handelt sich immer nur um verschiedene *Erscheinungsformen* ein und derselben Größe *Energie*". (Bergmann & Schaefer, 1998, p. 136)

In Westphal's *Physics*, 26th edition, energy is distinguished from work because energy describes a state, whereas work refers to an event in time.[23] According to Keller, Gettys and Skove (1993), work and heat are transferred energies:

> "*Heat* is energy transferred between a system and its environment because of a temperature difference between them [...] *Work* is energy transferred between a system and its environment by means independent of the temperature difference between them". (Keller et al., 1993, p. 423)

Cutnell and Johnson (1997) understand heat as energy in motion:

> "Definition of Heat
>
> Heat is energy that flows from a higher-temperature object to a lower-temperature object because of the difference in temperatures". (Cutnell & Johnson, 1997, p. 359)

According to Bergmann and Schaefer, energy and work are distinguished as follows: work is spoken of if energy is transported; energy is used if it is not in motion or is bound to a moving body.[24] Breithaupt states that energy can be transferred from one body to another by two methods, heat and work:

1. Work is energy transferred by means of a force moving its point of application.
2. Heat is energy transferred by means other than a force. A temperature difference is said to exist between two bodies if heat transfer between the two bodies could occur (Breithaupt, 1999, p. 376).

Tipler understands heat as energy that passes from one body to another.[25] Çengel and Boles define heat as a form of energy that is transferred between two systems.[26] These authors associate work and heat with processes and energy with state.[27] This has a consequence: "Systems possess energy, but not heat or work" (Çengel & Boles, 2002, p. 127). According to Serway, heat is the transfer of energy between a system and its environment due to a difference in temperature (Faughn et al., 2006, p. 353). Halliday, Resnick and Walker also understand work as energy that is transferred. They highlighted, however, that this transfer is not to be understood as something material that went from one body to another.[28] They explain the meaning of transfer

---

[23] "Energie beschreibt einen Zustand, Arbeit ist ein zeitlich ablaufender Vorgang". (Westphal, 1970, p. 39)

[24] "Der Begriff der Energie ist keineswegs nur ein anderes Wort für Arbeit. Erst wenn Energie in ganz bestimmter Weise transportiert wird, sprechen wir von Arbeit. Wenn sie dagegen ruht oder mit einem bewegten Körper fest verbunden ist, nennen wir sie schlicht Energie". (Bergmann & Schaefer, 1998, p. 136)

[25] "Wärme ist die Energie, die von einem Körper auf einen anderen aufgrund einer Temperaturdifferenz übergeht". (Tipler, 2000, p. 539)

[26] "Heat is defined as *the form of energy that is transferred between two systems (or a system and its surroundings) by virtue of a temperature difference*". (Çengel & Boles, 2002, p. 124)

[27] "Both are associated with a *process*, not a state". (ibid. p. 127)

[28] ""Arbeit" ist also Energie, die übertragen wird [...] Der Begriff "übertragen" kann auch irreführend sein. Er bedeutet nicht, dass etwas Materielles in das Objekt hinein oder aus dem Objekt herausfließt; diese Energieübertragung darf man sich also nicht wie fließendes Wasser vorstellen. Sie entspricht vielmehr dem elektronischen Geldtransfer zwischen zwei Bankkonten: Die Zahl in

through an electronic bank transfer. Therefore, according to this explanation, there is no real thing in motion.

5.2.2.1 Historical Connection

Lodge introduced the idea that energy is possessed by bodies and the aether and is transferred between them when there are events, such as the falling of a stone or the pendulum motion. Thus, we have energy at rest—possessed by bodies and the aether—and energy in motion—when it is transferred between bodies or between bodies and the aether. (This is the concept used by Preston.) As we have just seen above, some physicists have linked the term 'energy' with energy at rest (Lodge's possession of energy by bodies) and 'work' and 'heat' with energy in motion (Lodge's energy transfer). Halliday, Resnick and Walker also understand work as energy that is transferred but not in Lodge's sense.

5.2.3 Capacity of Doing Work

The *definition* of energy in the strict sense of the term definition that appears in textbooks tells us that energy is the capacity to do work or is the capacity that a body has to do work.

In Voigt's *Thermodynamics*, energy is understood as the body's capacity to do work.[29] Westphal defines energy as the capacity of doing work.[30] Breithaupt uses the two definitions mentioned above: first,

> "Energy is defined as the capacity to do work". (Breithaupt, 1999, p. 157)

and later in the book,

> "Energy is the capacity of a body to do work". (ibid. p. 376)

According to Tipler, the concept of energy describes the capacity to do work.[31] Serway and Beichner understand energy as the capacity that a body has for performing work (Serway & Beichner, 2000, p. 183).

This definition of energy has been criticized for decades.[32]

---

dem einen Konto steigt an, während die Zahl in dem anderen Konto kleiner wird, obwohl zwischen den beiden Konten kein materieller Gegenstand ausgetauscht wird". (Halliday et al., 2003, p. 154)

[29] "Entzieht man dem Körper die an ihm aufgewandte und in der ursprünglichen Form verschwundene Arbeit oder Wärme nicht in Form von Wärme oder Arbeit, so kommt sie eben der *Arbeitsfähigkeit*, der Energie des Körpers zu gute". (Voigt, 1903, p. 78)

[30] "Da Energie Arbeitsfähigkeit, also latente, aufgespeicherte Arbeit ist, so messen wir sie in den gleichen Einheiten wie die Arbeit". (Westphal, 1970, p. 39)

[31] "Der Begriff der Energie […] beschreibt die Fähigkeit, Arbeit zu verrichten". (Tipler, 2000, p. 129)

[32] Lehrman (1973), Sexl (1981), Duit (1981), Hicks (1983), Kemp (1984), Doménech et al. (2007), Hecht (2019). See also Appendix N.

5.2.3.1 Historical Connection

The definition of energy through 'capacity of doing work ' originates from Rankine. He wrote:

> "The term "*energy*" comprehends every state of a substance which constitutes a capacity for performing work". (Rankine, 1855, p. 217)

As Rankine's concept of substance means a body or a set of bodies, the term energy refers to any state of a body in which it has this capacity. In contemporary textbooks, it is usually stated that 'energy is the capacity to do work'. In this formulation, energy is the subject of the sentence, which may lead to the idea that energy has the capacity to do work. This wording, in turn, can lead one to think that energy is something real.

5.2.4 It Is not Known What Energy Is

In his *Lectures*, Feynman drew his students' attention to the following: "in physics today we have no knowledge of what energy *is*" (Feynman et al., 1963, Sect. 4.1). Keller, Gettys and Skove begin the chapter on the conservation of energy with three quotes where one is from Poincaré and one from Feynman. The latter has just been presented. The former tells us.

> "As we cannot give a general definition of energy, the principle of the conservation of energy simply signifies that there is something which remains constant". (Keller et al., 1993, p. 194)

In Bergmann and Schaefer's *Experimental Physics*, one reads "*nobody knows, what energy really is*". They add, "The physicist is almost in the same dilemma as the layman".[33] Hecht writes: "*there is no completely satisfactory definition of energy*" (Hecht, 2000, p. 222). Nevertheless, he points out "*energy is not an entity in and of itself – there is no such thing as pure energy*". (ibid. p. 223). Dransfeld, Kienle and Kalvius introduce the principle of energy conservation and ask the question "What actually is energy?" They answer: "we cannot answer it".[34] Çengel and Boles define thermodynamics as the science of energy.[35] However the authors point out the difficulty in defining energy:

> "Although everybody has a feeling of what energy is, it is difficult to give a precise definition for it. Energy can be viewed as the ability to cause changes". (Çengel & Boles, 2002, p. 2)

---

[33] "Definition der Energie. Dabei stoßen wir gleich auf eine Schwierigkeit: *Niemand weiß, was Energie wirklich ist.* Der Physiker befindet sich dabei fast im selben Dilemma wie der Laie". (Bergmann and Schaefer 1998, p. 135)

[34] "Hier stellt sich die Frage "Was ist eigentlich Energie?" Doch wir können sie, ähnlich der nach dem Wesen der Kraft, nicht beantworten". (Dransfeld et al., 2001, p. 109)

[35] "Thermodynamics can be defined as the science of *energy*". (Çengel and Boles 2002, p. 2)

Halliday, Resnick and Walker point out the difficulty in defining energy.[36]

5.2.4.1 Historical Connection

The prize competition of the Göttingen Philosophical Faculty, for which Planck wrote the book *The Principle of Conservation of Energy*, aimed at clarifying the concept of energy. It was precisely on this point that Planck failed. He gave no solution for how to understand the concept, but simply expected that the concept of his time would one day be overcome. Hertz was clearly saying that energetics was not advanced enough to define energy. Poincaré had two criticisms. He claims that it is impossible to give a definition of energy. He explains that this is impossible because we are unable to find a definition that encompasses all energy forms.[37] He also claims that the conservation principle holds, if one can sort energy into kinetic, potential and internal energy. If this sorting is not possible (he provides examples of this impossibility), the principle just tells us that something is conserved in nature. This is tautological, he concludes, because if there are laws of nature, something is conserved.[38]

5.3 Mayer's and Joule's Equivalents in Textbooks

Some textbook writers refer to Mayer and Joule. The reasons for mentioning Mayer are that heat is a form of energy, the principle of conservation of energy and the determination of the mechanical equivalent of heat.[39] The reasons for mentioning Joule are the mechanical equivalent of heat and his paddle wheel experiment, which is often illustrated. These illustrations differ, however, from each other.

---

[36] "In der Tat ist der Begriff der *Energie* so weit gefasst, dass eine einfache Definition nur sehr schwer zu geben ist. Zunächst einmal ist Energie eine skalare Größe, die mit dem Zustand eines oder mehrerer Objekte in Zusammenhang steht. Diese Definition ist jedoch zu vage, als dass sie eine echte Hilfe sein könnte". (Halliday, 2003, p. 152)

[37] "Nous connaissons beaucoup de formes de l'énergie, et rien ne nous prouve qu'il n'y en ait pas beaucoup d'autres qui nous sont inconnues. Dans chaque cas particulier, on voit bien ce que c'est l'énergie; mais il est impossible d'en trouver une définition générale". (Poincaré, 1901, p. 488)

[38] "Il ne nous reste plus qu'un énoncé pour le principe de la conservation de l'énergie; *il y a quelque chose qui demeure constant*. Sous cette forme, il se trouve à son tour hors des atteintes de l'expérience et se réduit à une sorte de tautologie". (Poincaré, 1897, p. 739)

[39] Preston (1919, p. 288 ff), Müller and Pouillet (1926, p. 110 ff), Hund (1956, pp. 50–51), Allen and Maxwell (1962, p. 284 ff), Arons (1965, p. 424 ff), McCormick (1969, p. 278 ff), Eisberg and Lerner (1983, pp. 406–407), Hecht (2000, pp. 540, 571), Cassiday et al., (2002, 255 ff), Tipler (2000, p. 554), Young and Freedman (2004, p. 653).

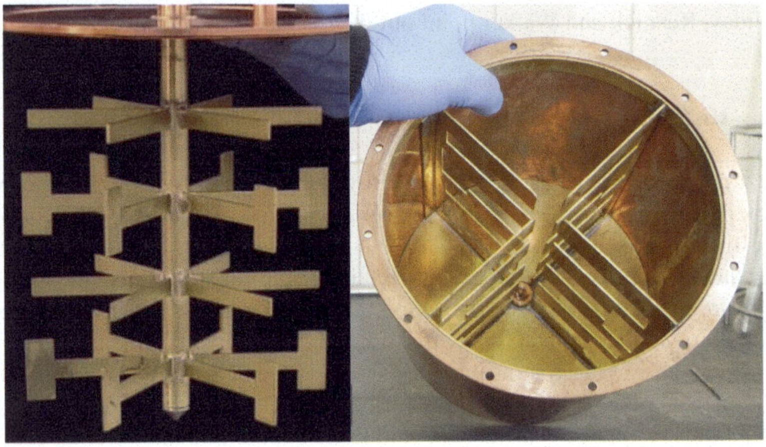

Fig. 5.1 Replica of the paddle wheel and the container in which the paddles rotate

5.3.1 The Paddle Wheel Experiment

Eisberg and Lerner (1983, p. 406) reproduced the original figure of the paddle wheel experiment (Joule, 1850). Hecht did not present the original figure, but his schema of the apparatus is conform to Joule's illustration (Figs. 2.14 and 2.15). Other illustrations differ from Joule's. To deal with this issue, I will start by showing a replica of Joule's, 1850 paddle wheel (Fig. 5.1).

In some illustrations in textbooks, the container appears without the internal obstacles. (Tipler & Mosca, 2004, p. 565; Serway & Beichner, 2000, p. 605) The paddles are also depicted as rectangular surfaces. The idea is that these surfaces push the water, when they rotate, and the water becomes warmer due to this motion. Now Joule's paddles are not that shape. These push the water but allow the water to collide with obstacles solidly attached to the walls of the container, as shown in the photo.

This begs the question, what happens if we remove these obstacles from the recipient? The difference is this. In Joule's machine, the bodies that set the paddle wheel in motion fall with constant velocity (Fig. 5.2). In the machine without obstacles, the falling motion is accelerated (Fig. 5.3). This has the following consequence. The weight of the bodies in the Joule machine when in motion is equal to the weight of the bodies at rest, i.e., before the phenomenon takes place. The weight of the bodies in the other machines when in motion, is less than the weight of the bodies at rest. Now, the magnitude 'weight times height' is decisive for the amount of heat that machines produce. Joule used the weight of bodies in the laboratory to calculate the mechanical equivalent of heat. This weight is equal to the weight of the bodies as they drive the rotation of the axis that sets the paddles in motion. In other machines, this is not the case; the weight in motion is lower. There is, therefore, a significant

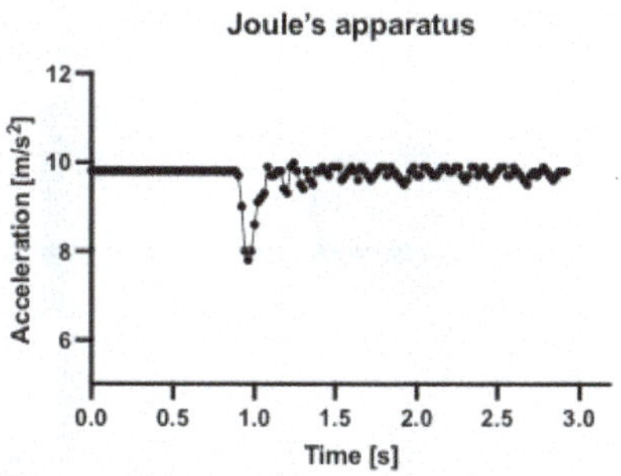

Fig. 5.2 This graph displays the result of a typical run of the experiment: the proper acceleration of the bodies as a function of time. The machine was kept at rest for about 0.8 s and released at some instant between 0.7 and 0.9 s. The experimental points in the interval ~ 0.8 to 3 s correspond to motion. The dispersion of the data points during motion is a characteristic trend observed in all runs of the experiment. Despite this behavior (also caused by oscillations of the axis of the apparatus), the average acceleration measured during motion by the accelerometer is conform with the acceleration at rest. The present and the following experiments used the accelerometers incorporated in the Pasco Scientific Smart Carts (ME-1240 and ME-1241). The data recorded by the accelerometer sensor is communicated to a PC, equipped with the *Capstone* software, by Bluetooth

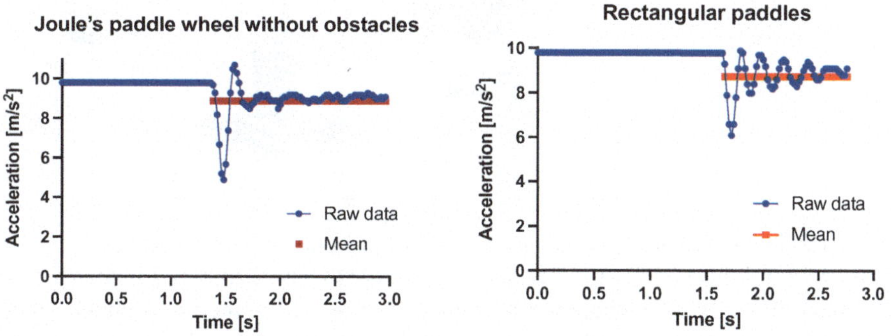

Fig. 5.3 In both cases, the container has no obstacles. The proper acceleration of these falling bodies is about 10% less than in the case of the original apparatus

difference between the Joule machine and those that have been presented in textbooks as a Joule machine.

5.3.1.1 Historical Connection

In Külp's *Experimental Physics*, Dreser presents the Joule machine in a simplified form (Fig. 5.4).

Verdet presents Joule's machine, the inside of the calorimeter (Fig. 5.5) and dealt with the experimental results in detail (Verdet, 1868, pp. 45–6).

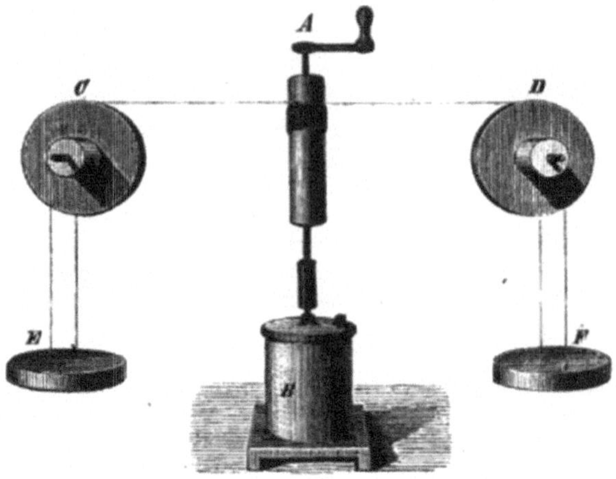

Fig. 5.4 Külp (1867, Vol. 4, p. 121)

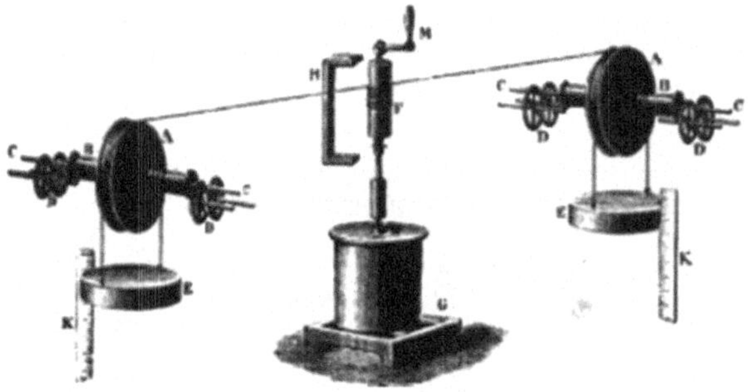

Fig. 5.5 Verdet (1868, p. 42)

Krebs' schema of Joule's machine (Fig. 5.6) presents only two paddles.

In 1878, Joule performed a new determination of the mechanical equivalent of the heat through the paddle wheel mechanism (Fig. 5.7). The container is smaller than in the 1850 experiment and the number of paddles is also reduced. The result is not significantly different from the 1850 result.

In 1879, Rowland performed a new determination of the mechanical equivalent of heat by means of a paddle wheel experiment. The idea is Joule's in the sense that the temperature difference of the water due to friction by the movement of a paddle wheel is measured. The main difference with regard to Joule's apparatus is that the friction has been optimized, i.e., the friction is not only due to the water hitting the internal parts (solidly connected to the container) but is also due to the water passing through small holes in these internal parts as well as in the paddles (Fig. 5.8). The value of the mechanical equivalent of heat is slightly higher than Joule's.

Fig. 5.6 Krebs (1877, p. 84)

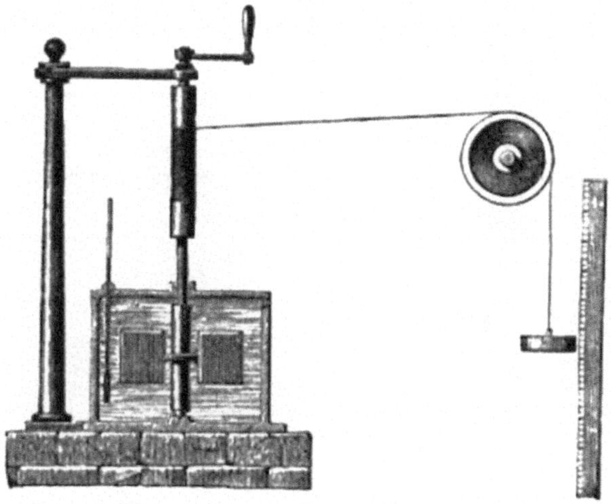

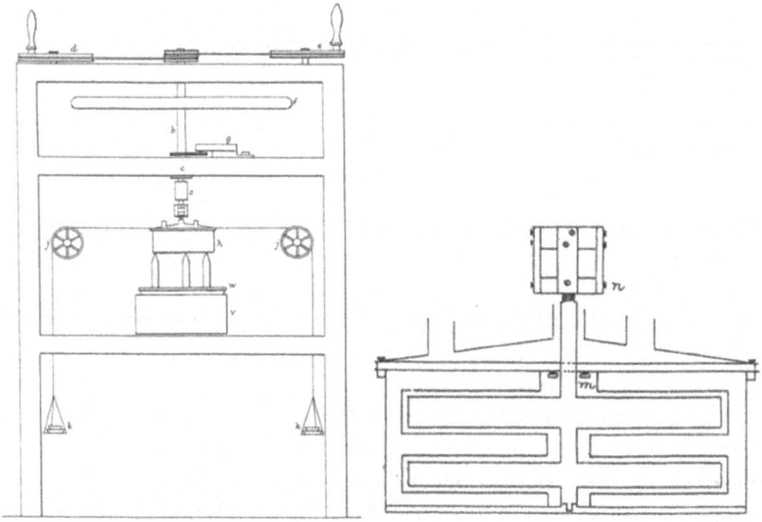

Fig. 5.7 A view of Joule's apparatus and of the inside of the container (Joule, 1878, Plate 26)

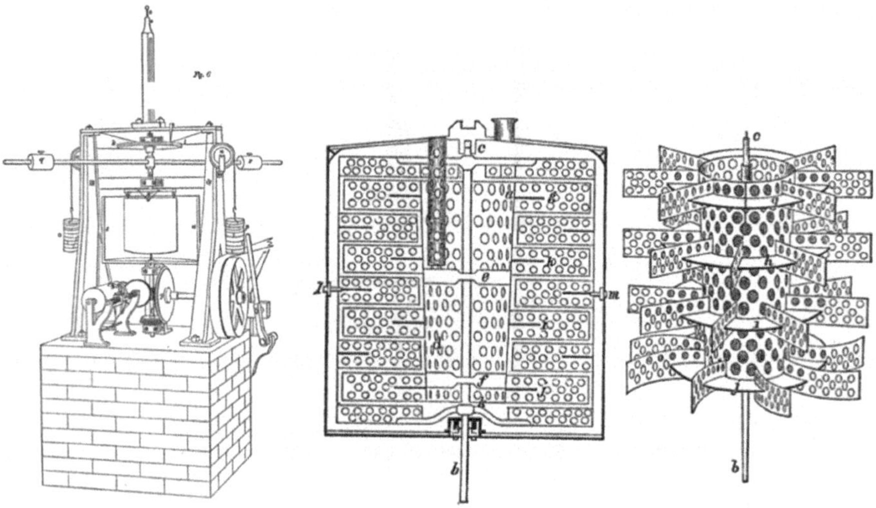

Fig. 5.8 A perspective view of the apparatus (Rowland, 1879, p. 159, Fig. 6). The other two pictures show a section of the calorimeter and a perspective view of the revolving paddles removed from the apparatus (ibid. p. 160).

5.4 Conclusion

Some physicists took the equivalence between heat and work as a principle, the principle of equivalence. This proposition consists of an induction: it generalizes an experimental result, the mechanical equivalent of heat. This principle has the particularity of expressing the first law of thermodynamics without using the concept of energy, which is why it was preferred by some physicists.

The impossibility of a *perpetuum mobile* dates back to a resolution of the Paris Academy of Sciences in 1775, which did not play any role in the idea of conservation in the 1840s. On the contrary, the assumption of conservation was taken as a (theoretical) justification for the impossibility of a *perpetuum mobile.*

The common formulation of the principle of conservation of energy goes back to Mayer's characteristics of force: indestructibility and transformability. The meaning of these characteristics has, however, changed in the course of time. Some physicists have taken energy to be an indestructible substance. Others, however, have highlighted that energy is not a substance. Such a divergence had already taken place in the nineteenth century.

The formulation of the first law of thermodynamic through the concept of internal energy of a body goes back to Thomson's mechanical energy. It was this concept that opened the door to the idea that energy exists in a body. The distinction between what concerns to the body and the surroundings, highlighted in textbooks, no longer comes from Thomson, but rather from Lodge. Contemporary textbooks take heat as transferred energy (Lodge's concept) and point out that the body has no heat, whereas Thomson admitted that a body has heat.

Depending on whether the authors use Mayer's properties of energy or Lodge's concept, they characterize heat as a form of energy or as energy transfer. Some authors use both ways, saying that heat is a form of energy and is transferred between systems. Beyond these concepts of energy, there is an explicit definition of it in some textbooks: energy is the ability to do work, which goes back to Rankine. Finally, there is a set of authors who argue that we do not know what energy is.

The overview of textbooks in this chapter raises some questions:

1. Some authors define energy, whereas others claim that we do not know what it is. Now, we might think that if we do not know what it is, we cannot define it. On the other hand, if the definition holds, there is no reason to say that we do not know what it is. It turns out that none of the textbook writers who claim that we do not know what energy is, show us where the definition of energy fails. Likewise, none of those who define energy show us that the others have missed something in the definition, which is why they do not accept it. This topic—energy is defined, on the one hand, and no clue to a definition, on the other—leads to the question of why the concept of energy leads to such contrary views.
2. Some authors teach that energy is neither created nor destroyed and understand that an entity with these characteristics exists. Other physicists deny that such a substance exists. We understand that at least one of these theses is false. What is the reason for such a divergence, which has persisted for over a century?
3. The common formulation of the principle of the conservation of energy and the formulation of the first law of thermodynamics through the concept of internal energy *are based on the concept of energy*. The principle of equivalence expresses the same relationship between heat and work without using 'energy'. One asks then the question of whether energy is really necessary.

These issues will be clarified in the next chapter.

References

Aktins, K. (1986). *Physik: die Grundlagen des physikalischen Weltbildes* (H.-W. Sichting, Trans.; 2 ed.). de Gruyter.

Allen, H. S., & Maxwell, R. S. (1962). *A text-book of heat*. Macmillan.

Arons, A. B. (1965). *Development of concepts of physics*. Addison-Wesley.

Benenson, W., Harris, J., Stocker, H., & Lutz, H. (2001). *Handbook of physics*. Springer.

Bergmann, L., & Schaefer, C. (1998). *Lehrbuch der Experimentalphysik* (11 ed., Vol. I). de Gruyter.

Böge, A., & Eichler, J. (2002). *Physik* (9 ed.). Vieweg.

Breithaupt, J. (1999). *Physics*. Macmillan.

Caneva, K. L. (2021). *Helmholtz and the conservation of energy: Contexts of creation and reception*. MIT Press.

Cassiday, D., Holton, G., & Rutherford, J. (2002). *Understanding physics*. Springer.

Çengel, Y., & Boles, M. (2002). *Thermodynamics*. Mc Graw Hill.

Çengel, Y., & Ghajar, A. (2015). *Heat and mass transfer: Fundamentals and applications* (5 ed.). Mc Graw-Hill Education.

Chabay, R., Sherwood, B., & Titus, A. (2019). A unified, contemporary approach to teaching energy in introductory physics. *American Journal of Physics, 87*, 504–509.

Clausius, R. (1879). *The mechanical theory of heat*. (W. R. Browne, Trans.). Macmillan.

Colding, L. A. (1856). Nogle Sætninger om Kræfterne (Theses Concerning Force). In P. Dahl (Ed.), *Ludvig colding and the conservation of energy principle*. Johnson Reprint Corporation.

Cutnell, J., & Johnson, K. (2006). *Essentials of physics*. J. Wiley.

Cutnell, J., & Johnson, K. (1997). *Physics*. Wiley.

Doménech, J. L., Gil-Pérez, D., Gras-Marti, A., Guisasola, J., Martínez-Torregrosa, J., Salinas, J., Trumper, R., Valdés, P., & Vilches, A. (2007). Teaching of energy issues: A debate proposal for a global reorientation. *Science & Education, 16*, 43–64.

Dransfeld, K., Kienle, P., & Kalvius, G. M. (2001). *Physik I: Mechanik und Wärme* (9 ed.). Oldenbourg.

Duit, R. (1981). Understanding energy as a conserved quantity—Remarks on the article by R. U. Sexl. *European Journal of Science Education, 3*, 291–294.

Eisberg, R. M., & Lerner, L. S. (1983). *Física: Fundamentos e Aplicações* (A. M. Costa, Trans.; Vol. 2). McGraw-Hill.

Elkana, Y. (1974). *Discovery of the conservation of energy*. Hutchinson.

Faughn, J. S., Serway, R. A., Vuille, C., & Bennett, C. A. (2006). *Serway's college physics* (7 ed.). Thomson Brooks/Cole.

Feynman, R. P., Leighton, R. B. & Sands, M. (1963). *The Feynman lectures on physics* (Vol. 1). Addison-Wesley.

Halliday, D., Resnick, R., & Walker, J. (1997). *Fundamentals of physics extended* (5 ed.). Wiley.

Halliday, D., Resnick, R., & Walker, J. (2003). *Physik* (S. Koch, Ed. & Trans.). Wiley.

Hänsel, H., & Neumann, W. (1993). *Physik: Mechanik und Wärme*. Spektrum, Akad. Verl.

Hecht, E. (2000). *Physics: Calculus*, (2nd ed., Vol. 1). Brooks/Cole.

Hecht, E. (2019). Understanding energy as a subtle concept: A model for teaching and learning energy. *American Journal of Physics, 87*, 495–503.

Helm, G. (1898). *Die Energetik nach der geschichtlichen Entwicklung*. Veit & C.

Hicks, N. (1983). Energy is the capacity to do work—Or is it? *Phys Teacher, 21*, 529–530.

Hudson, A., & Nelson, R. (1982). *University physics*. H. B. Jovanovich.

Hund, F. (1956). *Theoretische Physik* (Vol. 3). Teubner.

Joule, J. P. (1850). On the mechanical equivalent of heat. *Philosophical Transactions of the R s. of London, 140*, 61–82.

Joule, J. P. (1878). New determination of the mechanical equivalent of heat. *Philosophical Transactions of the R s. of London, 169*, 365–383.

Kane, J. W., & Sternheim, M. M. (1988). *Physics* (3 ed.). Wiley.

Keller, F. J., Gettys, W. E., & Skove, M. J. (1993). *Physics: Classical and modern* (2 ed.). McGraw-Hill.

Kemp, H. R. (1984). The concept of energy without heat and work. *Physics Education, 19*, 234–240.

Kipnis, N. (2014). Thermodynamics and mechanical equivalent of heat. *Science & Education, 23*, 2007–2044.

Krebs, G. (1877). *Die Erhaltung der Energie als Grundlage der neueren Physik*. K. Oldenbourg.

Külp, E. (1867). *Lehrbuch der Experimental-Physik* (Vol. 4). Diehl.

Lehrman, R. (1973). Energy is not the ability to do work. *American Journal of Physics, 60*, 356–365.

McCormick, W. W. (1969). *Fundamentals of university physics*. Macmillan.

Mayer, J. R. (1845). *Die organische Bewegung in ihrem Zusammenhange mit dem Stoffwechsel*. Drechsler.

Müller, J., & Pouillet, C. (1926). *Lehrbuch der Physik* (11 ed., Vol. 3-I). Braunschweig.

Nolting, W. (2002). *Theoretische Physik* (5 ed., Vol. 4). Vieweg.

Poincaré, H. (1892). *Cours de Physique Mathématique, 3. Thermodynamique: Leçons professés pendant le premier semestre 1888–89* (Vol. 3). J. Blondin.

Poincaré, H. (1897). Les idées de Hertz sur la mécanique. *Revue générale des Sciences, VIII*, 734–743.
Poincaré, H. (1901). *Sur les Principes de la Mécanique.* Ier Congrès International de Philosophie.
Preston, T. (1919). *The theory of heat* (R. Cotter, Ed., 3 ed.). Macmillan.
Rankine, W. (1855). Outlines of the science of energetics. *Edinburgh New Philosophical, 2*, 120–141.
Rowland, H. A. (1879). On the Mechanical equivalent of heat, with subsidiary researches on the variation of the mercurial from the air thermometer, and on the variation of the specific heat of water. *Proceedings of the American Academy of Arts and Sciences, 15*, 75–200.
Saha, M., & Srisvastava, D. (1935). *A treatise on heat* (2 ed.). The Indian Press.
Serway, R. A., & Beichner, R. J. (2000). *Physics for scientists and engineers* (5 ed.). Saunders College Publishing.
Sexl, R. U. (1981). Some observations concerning the teaching of the energy concept. *European Journal of Science Education, 3*, 285–289.
Smith, C. (1998). *The science of energy: A cultural history of energy physics in victorian britain.* The Athlone Press.
Tipler, P. A. (2000). *Physik.* Spektrum Akad. Verl.
Tipler, P. A., & Mosca, G. (2004). *Physics for scientists and engineers, extended version* (5th ed.). Freeman.
Verdet, E. (1868). *Oeuvres d'É. Verdet* (V. Prudhon, Ed., Vol. 7). Masson.
Voigt, W. (1903). *Thermodynamik* (Vol. I). G. J. Göschensche.
Westphal, W. (1970). *Physik* (26 ed.). Springer.
Wilson, J. D., & Buffa, A. J. (1997). *College physics* (3 ed.). Prentice Hall.
Wolf, F. (1949). *Grundzüge der Physik* (Vol. I). G. Braun.
Young, H., & Freedman, R. (2004). *Sears and Zemansky's university physics* (11 ed.). P. Addison-Wesley.

Chapter 6
Conclusion

The final chapter brings together the results of this study and provides the answer to the book's question, what is energy? The chapter begins with an overview of the evolution of the concept. The way in which the concept has changed over time and the diversity in contemporary times lead us to ask the question of whether there is something that cannot be changed, that is necessary to the concept. Two necessary conditions are then determined. Therefore, a concept that does not satisfy one of them cannot be an energy concept. In the next step, we arrive at an understanding of the concept of energy. Based on this, the reasons for the problems with the concept are explained. These problems that have arisen since the nineteenth century are thus overcome.

The caloric theory was the dominant theory until the mechanical equivalent of heat was accepted. Berthollet, Haldat, Carnot, Clapeyron, among others, argued that the amount of heat remains constant in nature. This idea was based on experimental work (Gay-Lussac, Berthollet, Dulong, Delaroche, Bérard, among many others). Rumford and Davy, however, argued that heat is motion. This concept of heat was based on experiments, in which the quantity of heat varies with motion (Table 6.1). The answer to the question of whether heat is a substance or a kind of motion depends, therefore, on the answer to the question of whether or not the *amount of heat is conserved* (Appendix A).

Joule showed experimentally that the quantity of heat does change and determined the mechanical equivalent of heat. He took this finding as a new argument against the heat-substance theory. He claimed then that heat is motion. This thesis about the nature of heat is an interpretation of his experimental result; as 'heat is a substance' was an interpretation of certain experiments. Indeed, neither Berthollet, Haldat, Carnot, Clapeyron observed the heat-substance nor Rumford, Davy, Mohr, Joule observed the motion of which heat should consist. Whereas Joule took a position about the nature of heat within the framework of the heat theory of the time, Mayer developed his own theory. Within the framework of this theory, heat is a form of force. Mayer was aware, however, that he was saying nothing about the nature of

R. Lopes Coelho, *What Is Energy?*, History of Physics,
https://doi.org/10.1007/978-3-031-51855-3_6

Table 6.1 Distinction between experimental results and their interpretation in key propositions

Experimental	Conceptual
The amount of heat is invariable	Heat is a substance
The quantity of heat is not invariable	Heat is motion

Table 6.2 The same experimental result and two interpretations

Experimental	Conceptual	Authors
Mechanical equivalent of heat	Heat is force	Mayer
Mechanical equivalent of heat	Heat is motion	Joule

heat. Heat being a form of force was, therefore, also an interpretation of phenomena (Table 6.2).

These two aspects provide us with an insight regarding the incompatibility between Carnot's theory and Joule's experiments. Thomson recognized the value of Joule's experiments but he did not accept his thesis on the nature of heat because it contradicted Carnot's axiom, heat is a substance (conceptual component). Clausius overcame this incompatibility. He argued that to eliminate the concept of heat as a substance, it is sufficient to prove that the amount of heat varies. (This concerns the experimental basis of the thesis 'heat is a substance'.) The variation of the heat quantity had been demonstrated by the experiments in which motion gives rise to heat, said Clausius based on Joule's experiments. Therefore, heat could not be a substance. From the dilemma of the time (either substance or motion), it followed that heat is motion. If heat is motion, the question of what kind of motion is heat, follows. Clausius was not interested in this question. (This concerns the conceptual component.) It was enough for him that heat could be thought of as motion [From this compatibilization arose the two laws of thermodynamics (Sect. 3.1)].

Thomson embraces the dynamic theory of heat in the following year. Within the framework of this theory, he introduces the term mechanical *energy* into the scientific vocabulary. He defines the mechanical energy of a body in a given state. This concept was not created to be applied to phenomena. It is, for example, meaningless to speak of 'energy transformation', energy being Thomson's mechanical energy. Thomson's concept refers to a pre-phenomenon situation, more exactly, what a body has as a reserve of work. It is, therefore, a concept that tells us how much work we can count on in a possible event. To account for this quantity, we need the mechanical equivalent of heat and to choose a standard state. This role of the equivalent is also clear in Thomson's mathematical expression of the mechanical energy (Sect. 3.2). We can see then that the discovery of the mechanical equivalent of heat was necessary to Thomson's mechanical energy.

At the same time, Thomson claims that a body loses work if it does work on other bodies (i.e., in the case that a phenomenon takes place) and that the quantity of work it loses is the same that the surroundings gain. Admitting this, Thomson presupposes that the quantity of mechanical energy considering the system body-surroundings remains constant. This means that the quantity of work in this system is conserved.

In sum, the term energy introduced by Thomson refers to the work reserve of a body, uses the mechanical equivalent of heat and presupposes the conservation of energy.

Based on Thomson's concept of mechanical energy, Rankine developed the concept of energy as the capacity of a body of doing work. This concept retains the original idea: energy is something that concerns a possible event. Rankine does more. He alters Thomson's stores of mechanical energy. A stone at a certain height from the ground represents a reserve of mechanical energy of the static kind, according to Thomson. With Rankine, it is potential energy. If the stone starts to fall, we no longer have the reserve of the static kind. With Rankine, the stone has potential energy throughout the whole motion. Moreover, it decreases during the motion in the same proportion as the other energy increases. In this way, Thomson's reserves, modified by Rankine, come to satisfy the conservation of energy. [This is the origin of the principle of conservation of energy in the form: the sum of the kinetic and potential energy is constant (Sects. 3.5–3.6)].

With this approach, we have now two types of energy forms, which are related to the two types of forms of force. Rankine's energies conform to Helmholtz's ultimate forces: tension force and living force. The living force concerns motion and the tension force concerns the tendency to motion. If one increases, the other decreases in the same proportion. This is Helmholtz's idea of the conservation of force (Sect. 2.4.1). According to Mayer (Sect. 2.1.2), there are five forms of force. Thus, in the 1840s we have a conservation with two forms of force (Helmholtz) and a conservation with five forms (Mayer) (Sect. 2.4.3.1). This situation reappears in the energy conceptual framework thanks to Rankine's modification of Thomson's kinds of mechanical energy. We found a consequence of this in Maxwell's *Theory of Heat*. He presents two formulations of the principle of conservation of energy: in one, the forms of energy are kinetic and potential (Rankine's forms); in the other, they are electric, magnetic, etc. (Mayer's forms). Regarding the former, Maxwell notes that we cannot guarantee that the forms are only these two, but, he adds, we cannot conceive of others (Sect. 3.6). Now we can understand this through the mechanical origin of those forms: one form concerns motion; the other includes rest (Thomson's dynamic and static kind of mechanical energy); we do not conceive of a third mechanical state beyond rest and motion. Hence, there is no third form of energy.

The potential character of Thomson's mechanical energy and Rankine's energy—the capacity of a body of doing work—is criticized by Lodge (Sect. 4.1.1). He argues then that energy is a real, existing thing. As we have seen, this real thing did not come from any experimental detection of it. One realizes, however, that his concept of energy can explain two things:

1. energy in a body
2. energy in an event.

The idea of energy in a body is justified by Lodge's claim that energy is possessed by bodies and by the aether. In order to account for the energy in an event serves the idea that energy is transferred between them. Let us continue. Something that can stay at rest (in a body or in the aether) and can move (between them) has the characteristics of a substance. Thus, we understand that energy was taken as a real existing thing. In sum, we understand that energy became a real existing thing to satisfy uses of the term 'energy':

- the use of energy with regard to an event, which comes from Mayer, and
- the use of the term with regard to a body, which comes from Thomson's mechanical energy of a body.

Lodge's reification of energy appears reinforced by Poynting's paper on the transfer of energy in the electromagnetic field, according to which energy moves through space. If energy moved through space, then the energy transfer thesis would be corroborated. Poynting himself pointed out, however, that there was no proof that energy moves through space.[1] Despite this, Lodge developed the idea that energy has an identity (Sect. 4.1.3). This identity of energy does not come, therefore, from experiments. His concept of energy satisfies, however, those requirements: to account for energy in a body and energy in events. Thus, we can conclude that Lodge's concept comes from his efforts in shaping a concept of energy in line with the uses of the term. (This concept is the basis of the common idea 'heat is transferred energy'.)

Planck's definition of energy generalizes the concept of mechanical equivalence (Sect. 4.2.1). This definition raises, however, the question of whether it is possible to obtain a mechanical equivalent with regard to any physical change. Planck proposes different means to achieve this goal and ends as follows: if it is not possible to express a physical change in mechanical terms in a certain case, the definition of energy is useless in that case. He was, however, able to overcome this difficulty through his first formulation of the principle of conservation of energy (Sect. 4.2.1). The problem for which he had no solution was the location of energy. If energy is constant in a system and energy is something real, it should be somewhere in this system, says Planck. It is not possible to pinpoint the place of energy. Hence, Planck predicted that this concept will one day be surpassed.

Starting in the 1860s, physicists appeared who used the concept 'principle of equivalence' to express the validity of the mechanical equivalent of heat.[2] Some of these physicists were reluctant to use the properties of energy expressed in the principle of conservation. Müller and Pouillet (1926) even pointed out that 'energy conservation' was a postulate, whose basis was the mechanical equivalent. For these authors, it was clear that the basis of the principle of conservation of energy was the mechanical equivalent of heat. Thus, one can understand their choice: to generalize the mechanical equivalent of heat, which was an experimental result, instead of resorting to a concept that was postulated.

What we have seen so far enables us to conclude the following. The mechanical equivalent of heat is the experimental basis of Mayer's theory, Joule's thesis,

---

[1] This proof is necessary, since the fact that an algorithm leads to correct predictions is not a sufficient reason to assert that the image based on it corresponds to reality. Indeed, different algorithms can predict the same experimental result and lead to different images of the phenomenon at stake (Coelho, 2021).

[2] Verdet (1868–72, p. 38 *et seq.*), Poincaré (1892, p. 53 *et seq.*), Preston (1919, p. 667), Müller and Pouillet (1926, p. 109), among others. See also Mayer (1851, pp. 41–3) and Caneva (1993, p. 260).

Thomson's definition of energy, Planck's definition of energy and the principle of equivalence. The indestructibility, transformability and immateriality of force (Mayer), heat as a state of vibration (Joule), the conservation of energy due to God (Thomson), the energy transfer between aether and bodies (Lodge) are elements of the theoretical constructions. This distinction between that which is experimental and that which is interpretation of experiments plays a heuristic role with regard to contemporary approaches.

The common formulation of the principle of conservation of energy states:

"energy can neither be created nor destroyed but only transformed".

This principle characterizes energy as uncreatable, indestructible and transformable. What is energy if based on these characteristics? Energy can be transformed. As transformation means change of form, it follows that what appears at the beginning and at the end of a transformation process are forms of energy. Only then can we say: one form is transformed into another. As these forms are observable (measurable), we can infer that the entity behind the forms is real. This being real is reinforced by the other properties of energy. If we are unable to destroy energy, it exists, since it does not make sense to be unable to destroy anything that does not exist. Furthermore, energy is independent of us since we cannot create it. Therefore, what we cannot destroy exists independently of us. The formulation of the energy principle presented above can lead us, therefore, to the concept of energy as a real existing thing.[3] This interpretation has no basis on the author who created these characteristics of energy.

Mayer's concepts of indestructibility and transformability refer to force and not energy. Nevertheless, the term 'force' used by Mayer refers to magnitudes that were and still are subsumed by the term energy.[4] Hence, it is legitimate to replace force by energy, where this reason holds. Substituting force by energy, we say, following Mayer, that energy is 'indestructible' and 'transformable'. Therefore, 'energy cannot be destroyed' and 'energy can be transformed', which is what appears in the energy conservation principle.

Indestructibility was abstracted from the following sequence (Table 6.3).

If the quantity of cause (A) is represented by A, of cause (B), by B and so on, we have that $A = B = C = \ldots$ From this sequence of equations Mayer concludes that *cause* is indestructible. Indestructible means, therefore, that the initial quantity holds along the chain. When he moves on to phenomena and finds that a *causal relationship* applies to a phenomenon, he uses the term force. Now he uses force for both members of the equation. Thus, the cause is called force and so is the effect

$$force(A) = force(B).$$

[3] Actually, studies have shown that contemporary textbooks take energy as a kind of a substance and students understand energy as a substance (Appendix N).

[4] Mayer's 'force of fall' corresponds to potential energy; the force called 'motion' to kinetic energy; heat, a form of force, has become a form of energy.

Table 6.3 Effect (B) equals cause (A) and becomes the cause of effect (C) and so on

Cause (A)	=	Effect (B)					
		↓					
		Cause (B)	=	Effect (C)			
				↓			
				Cause (C)	=	Effect (D)	
						↓	
						...	

With regard to 'cause = effect', he only changed the names. He had said that the cause was indestructible due to the sequence A = B = C = ... Then, he can also say that force is indestructible, what he did. He still stresses that force is "*quantitatively* indestructible" (Sect. 2.1.1). Therefore, force is indestructible regarding its quantity. There is, therefore, n*o entity that we cannot destroy*.

Transformability comes also from the same idea that cause equals effect and expresses a property for causes. This property is also ascribed to force. Approaching phenomena, Mayer comes to the causal relationship between heat and motion, assuming that heat is produced by motion or disappears to produce motion

$$\Delta heat = weight \times height$$

Is there a deep meaning of transformation? Mayer highlighted in different occasions that transformation has no ontological meaning (for example, Mayer, 1845, p. 10, 1851, p. 43). In the energy principle, however, 'transformability' contributes to the reality of energy in the way we have seen.

The third property of Mayer's energy is 'imponderability'. This property had a special meaning at that time. Light, heat, electricity and magnetism were called imponderables (Muncke, 1829; Suckow, 1813). Taking matter as ponderable and impenetrable, Mayer distinguishes it from force which is characterized by the opposite (Appendix D). Furthermore, Mayer pointed out that force is immaterial and stressed that there is no 'immaterial matter' (Mayer, 1845, p. 36). Therefore, Mayer's energy cannot be a substance.

In sum, Mayer's properties of energy—indestructible, transformable and immaterial—do not support the reading of the principle of conservation of energy that takes energy as something real, a kind of substance. (All this discussion concerns the conceptual component of the energy theory.)

Mayer arrived at the equation

$$0.000103\, thermal\ units = 3.77\, mechanical\ units \tag{6.1}$$

based on experimental data. From this equation, he calculated the mechanical equivalent of heat

$$1\, thermal\ unit = 36{,}602\, mechanical\ units \tag{6.2}$$

He called both sides of the equation 'force'. In this way, heat and motion, which were completely different from each other, became *homogeneous*. The conceptual homogenization was practiced by all four authors considered to be energy discoverers. Joule homogenized heat and motion through the concept of motion. He took heat as motion. Thus, instead of substance (heat) and motion, there are two kinds of motion. Both being motions, the equations of form

$$\alpha\, thermal\ units = \beta\, mechanical\ units$$

were justified. From these equations, he took the values of the mechanical equivalent of heat (Sects. 2.2.1–2.2.3). Colding homogenized heat and motion through the concept 'force of nature'. Heat and motion were forces of nature. As he admitted that forces of nature do not perish, he concluded that motion is transformed into heat (Sect. 2.3.1). Helmholtz homogenized heat and motion by means of the ultimate forces of nature: living force and tension force. Hence, these two forces exist in heat, as well as in electricity, electromagnetism, etc. (Sect. 2.4.1).

Let us go back to Mayer to continue our analysis of the energy concept. Mayer homogenized heat and motion by means of force. As he called 'force' both sides of the equation, he could say that force is indestructible. Since indestructibility concerns quantity, he could and did conclude that force remains constant (Sect. 2.1.2.1). The term 'force' was replaced by 'energy'. The properties of Mayer's force were transferred to energy. 'Energy' is then the *new name* given to both sides of the equation *cause* = *effect*. As one side of the equation is called energy and the other side also, it can be said for the same reason as before that energy is conserved. The authors who used the equivalence principle did not homogenize heat and work (Sect. 5.1.1). They had no concept that subsumes these. Therefore, they also lacked the corresponding quantity: the one that would encompass the quantities heat and work. They lacked, therefore, the quantity to be conserved. For this reason, they were even prevented from talking about *conservation*. Conclusion: we need to assign the same name to heat and work to have a principle of conservation. Without that, we do not have the quantity to conserve.

Mayer's axiom tells us that the quantity remains constant *in phenomena* ("In all physical and chemical phenomena" Sect. 2.1.2.1). What is the basis of this constancy? He said he stirred water vehemently and the temperature of the water rose. To find out whether the energy is conserved in this agitation of water, we need to measure the water temperature at the beginning and at the end of the agitation, determine the heat involved and the mechanical power employed. However, these values still do not tell us whether there was conservation. To know this, we need to know in advance which mechanical quantity corresponds to a unit of heat. Only then can we verify whether the measured values in the water agitation satisfy this proportion. Only if they satisfy, can we say there is conservation. Therefore, the mechanical equivalent of heat is a *conditio sine qua non* of energy conservation.

Conclusion: from what we have just seen and from what we saw before, it follows that to say that energy is conserved we need two things:

1. that the mechanical equivalent of heat is valid
2. that we give the same name 'energy' to both sides of the equation.

The mechanical equivalent of heat is also the basis of the principle of equivalence. So, I will use it to clarify what is at stake:

$$1\,\text{cal} = 4.186\,\text{J} \tag{6.3}$$

The equivalence principle uses this result by focusing on the equality. This equivalence is not only valid for a small number of types of experiments but is general. This generality is expressed by the term 'principle' in 'principle of equivalence' (This is how Verdet and Poincaré demonstrate the principle, Sect. 5.1.1.1). An alternative to this approach is to homogenize the members of the equation. Doing so using "energy", 1 cal is a quantity of energy and 4186 J is a quantity of energy.

$$\begin{array}{ccc} 1\,\text{cal} & = & 4.186\,\text{J} \\ \downarrow & & \downarrow \\ \text{E} & & \text{E} \end{array}$$

The equality of the quantities of energy is guaranteed from the beginning by the initial Eq. (6.3). It then follows, going on to the phenomena, that the quantity referring to a given aspect of the phenomenon (heat) is equal to the quantity referring to the other aspect (work). Since the quantity in the first phase of the phenomenon is equal to the quantity in the other phase of the phenomenon, we say that the quantity is conserved in that phenomenon. As this is true in general, 'principle' is used in 'principle of conservation of energy'. What does this principle say? If we want to say what this principle tells us from what we have said so far, then we say: this principle tells us that the quantity of energy is conserved in phenomena. (This is Mayer's formulation of the principle (Sect. 2.1.2.1). It raises no ontological problems, as we have seen.)

When it was assumed that energy is indestructible, uncreatable and transformable, the justification of the principle shifted to this entity, as seen above (see also Sect. 2.1.2.1). In fact, an entity with these characteristics—it always exists in the same quantity and adapts to what we observe (due to transformability)—forces us to accept the principle in the sense that we realize that it must be true. That this entity 'energy' was the consequence of a substantialization of energy, (Sects. 4.1 and 4.3) had already been forgotten (Sect. 5.1.1). That this substantialization was possible because the equivalence (6.3) was homogenized by the term energy, had not been borne in mind. (As we have seen, without the use of the same name 'energy' in both members of the equivalence, there is no conserved quantity.) Thus, the principle of conservation of energy, which originally said nothing more than that—the quantity of energy is constant in natural phenomena—began to indicate properties of energy that justify the principle itself. When the question was asked, what is this that is neither created nor destroyed, difficulties arose that have lasted for more than a century (Chaps. 1, 4 and 5).

Synoptically, we have the following:

1. mechanical equivalent of heat (this was the foundation of the theories of Mayer, Joule, Thomson, etc.)
2. Homogenization (assigning the same name to both members of the equivalence. If the members have the same name, the quantities are the same.)
3. Formulation of the principle: the quantity of the homogenization remains constant (if this quantity is energy, then the principle states: the quantity of energy is conserved)
4. Energy is a substance (the result of reification processes)
5. Contemporary formulation of the principle: energy is neither created nor destroyed, but only transformed or transferred.

About 1 and 2: The mechanical equivalent of heat does not imply homogenization, because the equivalence principle expresses the equivalent without homogenization (Sect. 5.1.1). Originally, the first law of thermodynamics also expressed the mechanical equivalent of heat without homogenizing (Sect. 3.1). From this, we understand that homogenization is not necessary, but it is an option.

About 2: Homogenization was achieved in various ways: Mayer, by force; Joule, by motion, etc., as seen above. It follows that the means of homogenization is also an option.

About 3: The quantity resulting from homogenization remaining constant expresses the principle of conservation in Mayer's formulation (Sect. 2.1.2.1).

About 4: For energy to be a substance, the quantity had to be conserved. Without conservation of quantity, there would be no substance. (This is analogous to what happened with heat. Heat was a substance because the quantity was constant; when the quantity was found to vary, the thesis collapsed.)

About 5: This formulation of the principle of conservation of energy dispenses with the mechanical equivalent of heat as a basis. The latter becomes a consequence of the existence of energy.

Understanding the processes of homogenization and reification of energy allows us to overcome the problems posed by physicists since the nineteenth century.

Planck's problem consisted of the following. Since the energy of an isolated system remains constant, the energy should be somewhere in the system, as it was understood as a substance. However, it is impossible to localise the energy in the system (Sect. 4.2.1).

In fact, the energy-substance was never located. Now we know why it was not. When we go to the origin (Mayer's force), we realize that force (later energy) was the name given to the two members of the equivalence. 'Energy' is a homogenizing term. (It is one of these terms since other homogenizations have been carried out.) Moreover, homogenization itself, whatever term is used, is not necessarily imposed, as the equivalence (6.3) can be interpreted without it (Sect. 5.1.1). Thus, we realize that energy is a conceptual product, not part of nature. It is therefore understandable that it is impossible to locate it in a system where energy is conserved, which was Planck's problem.

Poincaré defended that "it is impossible to find a general definition" of energy, for the following reason (Chap. 1). In each case of energy transformation, it was well known which energy forms were involved. Coming up with a concept of energy that encompassed all forms was, however, impossible (Chap. 1, Sect. 5.2.4.1). This conclusion provides an answer to the following question: from the phenomena in which the forms of energy are identified, can I arrive at what is common to all forms? If so, that would be energy. The question makes sense but has no solution. The reason is this. The 'forms' were created to overcome a difficulty in the theory itself. Mayer said that cause is equal to effect. The cause could be heat and the effect motion. However, heat and motion are different. Therefore, to say that they were the same was contrary to what was observed. With the transformability of force, the problem has been overcome: the core is the same (force/energy), with only one variation (the form is different). Therefore, "energy form" is not something that is abstracted from the phenomenon.

When energy became a real, existing thing, forms of energy began to be understood as manifestations of this reality. Therefore, it would have made sense to study the forms of energy to understand what energy is. It turns out, however, that energy has been substantialized intellectually (Sects. 4.1 and 4.3). Consequently, energy as a substance only exists as a concept. Therefore, the forms of energy in phenomena—potential, kinetic, thermal, electrical, etc.—cannot lead us to the core of what they would be a manifestation of, because that core does not exist. Thus, Poincaré's conclusion is understandable: based on the energy forms, it is impossible to arrive at a definition of energy.

Feynman points out that in physics today we have no knowledge of what energy *is*' (Chap. 1, Sect. 5.2.4). This statement presupposes that there should be knowledge of what energy is by means of physics. But are the means of physics adequate to achieve this knowledge? Bergmann and Schaefer said that the physicist is almost in the same dilemma as the layman when it comes to the question of what energy is (Sect. 5.2.4). This shows that these authors could not find the appropriate means in physics to deal with this question. In fact, the determination of the equivalent is a physical issue but the interpretation of the members of the equivalence that emerged in the nineteenth century is no longer. The process of substantializing energy that emerged at the end of the nineteenth century is not the subject of any chapter in physics. The fact that the principle of conservation of energy was replaced by the equivalence principle by some physicists and that the latter represents, therefore, an alternative to the former, is not known in physics today. If what could have led us to an understanding of energy is not part of physics, it is understandable that there is no knowledge in physics of what energy is, as Feynman said.

Other physicists pointed out the same difficulty (Sect. 5.2.4) or criticized the concept. However, most physicists define energy and formulate the principle of energy conservation, as if energy existed (Sects. 5.2.1 and 5.2.2). For example, Dransfeld et al. wrote: "Energy can be neither produced nor destroyed but remains constant in every closed system. (Fact of experience!)" (Dransfeld, 2001, p. 109, Sect. 5.1.3).

The first part of the proposition presupposes homogenization, as seen above. The fact that we give the same name of energy to the two members of an equivalence is not a "fact of experience" (I recall that according to Müller and Pouillet the conservation of energy was a postulate, not a fact of experience, Sect. 5.1.1). Therefore, the first part of Dransfeld's proposition is not a fact of experience. We know, however, that the principle of energy conservation is based on solid experimental work (this is the reason why physicists defend the principle). Taking this experimental work into account, we might think that we have found a justification for the principle. In doing so, however, I am resorting to something that is not in the proposition. What the proposition says is that the fact of experience is this: energy is neither created nor destroyed. When the question is then asked, what experiment shows me that an entity with these characteristics exists, difficulties arise. Thus, we understand why physicists have defended the principle of energy conservation and its formulation has led to the problem with the concept of energy.

References

Caneva, K. L. (1993). *Robert Mayer and the conservation of energy*. Princeton University Press.

Coelho, R. L. (2021). On the composition of force: Algorithm and experiment. *Axiomathes, 31*, 199–210.

Dransfeld, K., Kienle, P., & Kalvius, G. M. (2001). *Physik I: Mechanik und Wärme* (9 ed.). Oldenbourg.

Mayer, J. R. (1845). *Die organische Bewegung in ihrem Zusammenhange mit dem Stoffwechsel. Ein Beitrag zur Naturkunde*. Drechsler.

Mayer, J. R. (1851). *Bemerkungen über das mechanische Aequivalent der Wärme*. Landherr.

Müller, J., & Pouillet, C. (1926). *Lehrbuch der Physik* (11 ed., Vol. 3-I). Vieweg.

Muncke, G. W. (1829). *Handbuch der Naturlehre* (Vol. I). Universitäts-Buchhandlung C.

Poincaré, H. (1892). *Cours de Physique Mathématique, 3. Thermodynamique: Leçons professés pendant le premier semestre 1888–89* (Vol. 3). J. Blondin.

Preston, T. (1919). *The Theory of Heat* (R. Cotter, Ed., 3 ed.). Macmillan.

Suckow, G. A. (1813). *Anfangsgründe der Physik und Chemie nach den neuesten Entdeckungen*. Augsburg.

Verdet, E. (1868). *Oeuvres d'É. Verdet* (V. Prudhon, Ed., Vol. 7). Masson.

Appendix A
Heat: Either Substance or Motion

In the late eighteenth century and the first half of the next, there were two conflicting theses on the nature of heat: 'heat is a substance' and 'heat is motion'. For some authors, these theses were only hypotheses.

Laplace and Lavoisier (1780) account for divergences on the nature of heat. There are authors who take heat as a fluid that penetrates bodies, they say.[1] For others, heat results from unobservable movements of "molecules of matter".[2] In this case, they add, heat would consist of the sum of the living force of the molecules of matter.[3] Lavoisier and Laplace decided then to opt for neither of those two hypotheses[4] and added: "In the ignorance where we are on the nature of heat, it only remains for us to observe its effects carefully".[5]

Lavoisier, Morveau, Berthollet and Fourcroy had carried out a reform of the chemical nomenclature, in which the element 'caloric' appears. By "caloric" was meant a subtle fluid that lies between molecules and is responsible for our sensation of heat.[6] This could lead one to think that Lavoisier had adopted the hypothesis of heat being a substance. In the *Traité Élémentaire de Chimie*, he explains, however, that he had adopted the term 'caloric' because he thought of heat as being the effect

---

[1] "Les physiciens sont partagés sur la nature de la chaleur. Plusieurs d'entre eux la regardent comme un fluide répandu dans toute la nature, et dont les corps sont plus ou moins pénétrés". (Laplace & Lavoisier, 1780, p. 10)

[2] "D'autres physiciens pensent que la chaleur n'est que le résultat des mouvements insensibles des molécules de la matière". (ibid. p. 10)

[3] "la chaleur est la force vive qui résulte des mouvements insensibles des molécules d'un corps; elle est la somme des produits de la masse de chaque molécule par le carré de sa Vitesse". (ibid. p. 11)

[4] "Nous ne déciderons point entre les deux hypothèses précédentes". (ibid. p. 12)

[5] "Dans l'ignorance où nous sommes sur la nature de la chaleur, il ne nous reste qu'à bien observer ses effets". (ibid. p. 14)

[6] "Il est difficile de concevoir ces phénomènes sans admettre qu'ils son l'effet d'une substance réelle et matérielle, d'un fluide très-subtil". (ibid. p. 19)

R. Lopes Coelho, *What Is Energy?*, History of Physics,
https://doi.org/10.1007/978-3-031-51855-3

of something and wanted to distinguish between cause and effect.[7] To make this distinction, he used 'caloric' to denote the cause. He adds, however, that this concept has the advantage of adapting to any opinion, because caloric is not necessarily an actual matter.[8] What is it then? It is a name that refers to the repulsive cause, whatever it may be, that drives molecules away from each other. (In this conceptual framework, the state of bodies resulted from a balance between two forces: the attractive one, which came from gravitation, and the repulsive one, which was then due to caloric.[9])

Pictet (1790) also gives an account of the divergence among physicists about the nature of heat. Some take heat to be a vibration of the particles that constitute bodies. These attribute the change in temperature to the variation in the intensity of the vibrations.[10] For others, heat is a *sui generis* fluid that penetrates bodies and dilates them.[11] Pictet foresaw that the designation 'caloric' would be generally adopted.[12]

Black's lectures were published posthumously, in 1803. He taught that the theses 'heat is substance' and 'heat is motion' are hypotheses. The fact that they prove adequate to phenomena is not a decisive argument because for Black, it is possible to adapt any theory to the phenomena.[13] He goes one step further: he highlights the reasoning behind these hypotheses.

He argues, for example, that Lord Verulam had concluded that heat is motion[14] because he had obtained heat by means of motion.

---

[7] "Cette substance, quelle qu'elle soit, étant la cause de la chaleur, ou, en d'autres termes, la sensation que nous appelons chaleur [...] on ne peut pas, dans un langage rigoureux, la désigner par le nom de chaleur, parce que la même dénomination ne peut pas exprimer la cause et l'effet". (ibid. p. 19)

[8] "Nous [Morveau, Berthollet, Fourcroy, Lavoisier] avons en conséquence désigné la cause de la chaleur, le fluide éminemment élastique qui la produit, par le nom de *calorique* [...] elle [cette expression] a encore un autre avantage, c'est de pouvoir s'adapter à toutes sortes d'opinions; puisque, rigoureusement parlant, nous ne sommes pas même obligés de supposer que la calorique soit une matière réelle". (ibid. p. 19)

[9] "les molécules des corps peuvent être considérées como obéissant à deux forces, l'une répulsive, l'autre attractive, entre lesquelles eles sont en équilibre". (Lavoisier, 1864, p. 18)

[10] "les plus grands Physiciens de notre siècle ne sont point encore d'accord sur la nature du feu; les uns le regardant comme une simple modification des corps, comme un movement de vibration des particules qui composent les aggrégés visibles, & ils attribuent les changements de température aux variations dans l'intensité de ces vibrations". (Pictet, 1790, p. 3)

[11] "le feu selon d'autres Physiciens est un fluide particulier, *sui generis* qui passe aisément au travers de tous les corps, qui est prodigieusement expansible, & qui, par une suite de son élasticité ou par quelqu'autre cause, dilate tous les corps qu'il pénétre". (ibid. p. 3)

[12] "plusieurs Ecrivains, soit dans notre langue, soit dans la langue anglaise, désignent la cause sous le nom qui auroit dû appartenir exclusivement à l'effet; ils appellent *chaleur* & quelquefois *matière de la chaleur* le feu lui-même considéré dans son état de liberté; ou l'a encore appelé *fluide igné*, *fluide calorifique*; enfin les Chymistes célèbres, qui viennent de proposer une nouvelle nomenclature, le nomment *calorique*, expression heureuse, & qui sera sans doute généralement adoptee". (ibid. p. 5)

[13] "A nice adaptation of conditions will make almost any hypothesis agree with the phenomena". (ibid. p. 193)

[14] "The only conclusion, however, that he [Lord Verulam] is able to draw from the whole of his facts, is a very general one, viz. that heat is motion". (ibid. p. 31)

> "This [his] conclusion is founded chiefly on the consideration of several means by which heat is produced, or made appear, in bodies; as the percussion of iron, the friction of solid bodies, the collision of flint and steel". (Black, 1803, p. 31)

It is true that motion is needed to strike the iron and the iron gets hotter. But between this motion and the appearance of heat there is the striking of the iron, which is neglected in that interpretation of the phenomenon.

Regarding the explanation about the communication of heat from the warm body to the cold body,[15] Black points out the implicit analogy between the ordinary experience of communication of matter and motion and the communication of heat:

> "The vulgar opinion is, that the hot body has lost something which has been added to the other. And those who have attempted to reason more profoundly on the nature of heat, have agreed with the multitude in this point; and have supposed that heat is a positive quality, and depends, either upon an exceedingly subtile and active matter, introduced into the pores of bodies, or upon a tremor or vibration excited among their particles, or perhaps among the particles of a peculiar substance present in all bodies; which subtile matter, or tremulous motion, they have supposed to be communicated from the hot body to the colder, agreeably to our general experience of the communication of matter or of motion". (ibid. p. 25)

This explanation of the communication of heat has a premise: the warm body gives something to the cold body. One could also assume that the cold body is active. In that case, the cold body would give something to the warm body. This hypothesis also existed. Frigorific particles that came from the cold body were admitted.[16]

Under these conditions, one wonders how to proceed in the study of phenomena. Black answers: "Our first business must, therefore, necessarily be, to *study the facts* belonging to our subject" (ibid. p. 35).

Rumford (1798) and Davy (1799) argued that heat was motion. Both conducted friction experiments. By rubbing two bodies together, they found that the amount of heat increased. If the amount of heat increased, heat could not be a substance.

According to Haldat (1807), physicists were divided between two diametrically opposed theories held by reputable authorities. This is Rumford, who argued that heat was a vibratory movement analogous to sound and Berthollet, who had developed a refutation of Rumford's thesis.[17] Haldat conducts friction experiments. His research concerns the relationship of the heat developed with different materials,

---

[15] "When we consider this communication of heat from hot bodies to colder ones, the first question which may naturally occur to our mind, is, In what manner have these two bodies acted, the one on the other, on this occasion? Has one of them lost something, which the other has gained? And which of them has lost, or which has received?". (ibid. p. 25)

[16] "they have supposed that there is in the ice, or cold iron, a multitude of minute particles, which they call particles of frost, or frigoric particles, and which have a tendency to pass from the very cold bodies into any others that are less cold". (ibid. p. 26)

[17] "On connoissoit depuis long-temps la propriété qu'a le frottement de développer la chaleur, mais on n'avoit encore soumis à aucun examen approfondi ce fait si digne d'attention. M de Rumford [...] qu'elle avoit été produite par l'agitation des molécules communiquées à l'eau à la manière du son. Cependant cette conclusion, qui tend à renverser de fond en comble la théorie du calorique, n'a pas paru légitimement déduite des faits, et M. Berthollet l'a réfutée dans une note de sa Statistique chimique [...] L'opinion des physiciens demeure ainsi flottante entre deux théories diamétralement opposées et soustenues l'une et l'autre par des autorités respectables". (Haldat, 1807, p. 214)

taking into account their density and the roughness of the surfaces. Regarding the primary question he asked in the paper, the nature of heat, he concluded that the experiments did not lead to an answer.[18] Thus, on one hand, Haldat corroborates Berthollet's thesis, the causes of frictional heat arise from the compression of the molecules of the bodies.[19] On the other hand, he says, there are facts that seem to corroborate Rumford's opinion.[20] Nevertheless, since there are so many arguments in favor of the materiality of heat, Haldat chooses to maintain the theory until further clarification becomes available.[21]

In the same year, Young defends a different thesis: "If heat is not a substance, it must be a quality; and this quality can only be motion" (Young, 1807, Lecture LII, p. 654).

J. T. Mayer (1820) argues that there is no such thing as heat as a substance. Heat comes from the internal motion of bodies. He does not specify the type of internal motion but takes it as an effect of attractive and repulsive forces.[22]

Colladon and Sturm (1828) perform compression experiments. They also point out that the appearance of the heat of compression is connected with the most important questions of physics and could lead to interesting conclusions about the nature of heat.[23]

Kämtz argues that there are two *hypotheses* about the nature of heat: heat as a substance and as motion (Kämtz, 1839, pp. 428–9).

Matteuci uses the term 'caloric' to designate the cause or force which has the effects we call heat. He points out, however, that we ignore the nature of this force or the cause of heat (Matteuci, 1842, p. 1).

---

[18] "Si les expériences que je viens de décrire n'ont pas eu tout le succès que j'en attendois pour déterminer la cause productrice de la chaleur qui se dégage dans le frottement des corps, elles ne sont pas cependant dépourvues d'utilité, ce me semble [...]". (ibid. p. 220)

[19] "De toutes les causes soupçonnées d'influer sur la production de la chaleur dans nos expériences, aucune ne paroît plus puissante que la condensation des molécules du corps résultant de la pression nécessaire pour opérer le frottement. C'est aussi à cette cause que M. Berthollet a cru devoir l'attribuer; mais comme son influence n'a été déterminée que par le raisonnement, j'ai cherché à la constater par l'expérience". (ibid. p. 219)

[20] "Ces faits opposés au résultat qu'il y avoit lieu d'attendre, rendent très difficile l'explication des phénomènes calorifiques produits par le frottement. Ils sembleroient favoriser l'opinion de M. de Rumford". (ibid. p. 221)

[21] "Cependant un si grand nombre d'argumens établissent tellement la matérialité du calorique, que l'on ne doit pas abandonner une théorie si féconde en explications utiles, avant que de nouvelles expériences aient éclairci totalement ce point important de doctrine". (ibid. p. 221)

[22] "Es gibt gar keine solche eigene von der Materie eines Körpers selbst verschiedene Wärmematerie (Wärmestoff), sondern die Wärme rührt nur von einem gewissen Zustande der Körper, einer gewissen innern Bewegung ihrer Theile selbst her [...] Oder endlich ist das Gefühl von Wärme nur der Erfolg eines geänderten Verhältnisses der Attractio- und Repulsivkräfte". (Mayer, 1820, p. 233)

[23] "Die bei der Zusammendrückung der Körper auftretenden Wärmeerscheinungen haben seit einigen Jahren die Aufmerksamkeit mehrerer Mathematiker und Physiker auf sich gezogen. Die Kenntniß dieser Erscheinungen schließt sich an die wichtigsten Fragen in der Physik, und kann vielleicht zu höchst interessanten Folgerungen über die Natur der Wärme selbst und über die zwischen ihr und den verschiedenen Körpern stattfindenden Beziehungen führen". (Colladon & Sturm, 1828, p. 161)

Conclusion:

- Heat is a substance, according to Karsten (1790), Haldat (1807), Carnot (1824), Clapeyron (1834) Thomson (1848, 1849), among others,
- Heat is motion, according to Rumford (1798), Davy (1799), Young (1807), J. T. Mayer (1820), Mohr (1837), Joule (1843), Helmholtz (1882), Clausius (1850), Thomson (1851), among others,
- These two theses are only hypotheses, according to Laplace and Lavoisier (1780), Black (1803), Kämtz (1839).

With the acceptance of the mechanical equivalent of heat, the thesis of heat as a substance fell away. Heat as motion remained. Some authors, however, shy away from the question about the nature of heat. As we have seen, the principle of equivalence had precisely the advantage for some physicists that it was not connected with the nature of heat.

References

Black, J. (1803). *Lectures on the Elements of Chemistry, Delivered in the University of Edinburgh* (J. Robison, Ed. Vol. 1). Creech.

Carnot, S. (1824). *Réflexions sur la puissance motrice du feu.* Bachelier.

Clapeyron, B. (1834). Mémoire sur la puissance motrice de la chaleur. *Journal de l'École Polytechnique, XIV*, 153–190.

Clausius, R. (1850). Ueber die bewegende Kraft der Wärme und die Gesetze, welche sich daraus für die Wärmelehre selbst ableiten lassen. *Annalen der Physik, 79*, 368–397, 500–524.

Colladon, J.-D., Sturm, C. F. (1828). Ueber die Zusammendrückbarkeit der Flüssigkeiten. *Annalen der Physik, 88*, 161–197.

Davy, H. (1799). An Essay on Heat, Light, and Combinations of Light. In J. Davy (Ed.), *Collected Works* (Vol. 2, pp. 2–86.). Smith, Elder and Co.

Haldat. (1807). Recherches sur la chaleur produite par le frottement. *Journal de Physique, de Chimie et d'Histoire Naturelle, 65*, 213–222.

Helmholtz, H. (1882). *Wissenschaftliche Abhandlungen* (Vol. I). Barth.

Joule, J. P. (1843). On the Calorific Effects of Magneto-Electricity, and on the Mechanical Value of Heat. *The London, Edinburgh, and Dublin Philosophical Magazine and Journal of Science, Vol. XXIII*, 263–276, 347–355, 435–443.

Kämtz, L. F. (1839). *Lehrbuch der Experimentalphysik.* Halle: Gebauer.

Karsten, W. J. G. (1790). *Anfangsgründe der Naturlehre.* Halle: Rengersche Buchhandlung.

Laplace, P. S., Lavoisier, A. (1780). Mémoire sur la chaleur. In *Oeuvres Complètes de Lavoisier* (Vol. 10, pp. 149–200). Gauthier-Villars.

Lavoisier, A. L. (1864). Traité élémentaire de Chimie. In *Oeuvres Complètes de Lavoisier* (3 ed., Vol. I). Paris.

Matteuci, C. (1842). *Lezioni di Fisica* (Vol. 3). Nistri.

Mayer, J. T. (1820). *Anfangsgründe der Naturlehre* (4 ed.). Dieterichsche Buchhandlung.

Mohr, F. (1837). Ueber die Natur der Wärme. *Zeitschrift für Physik und verwandten Wissenschaften, V*, 419–445.

Pictet, M.-A. (1790). *Essais de Physique.* Barde, Manget et Cie.

Rumford, B. C. O. (1798). An inquiry concerning the Source of the Heat which is excited by Friction. *Philosophical Transactions of the Royal Society of London, 88*, 80–102.

Thomson, W. (1848). On an Absolute Thermometric Scale founded on Carnot's Theory of the Motive Power of Heat. *Philosophical Magazine*, *33*, 313–317.

Thomson, W. (1849). An Account of Carnot's Theory of the Motive Power of Heat; with Numerical Results deduced from Regnault's Experiments of Steam. *Transactions of the R. S. of Edinburgh*, *16*, 541–574.

Thomson, W. (1851). On the Dynamical Theory of Heat; with numerical results deduced from Mr Joule's Equivalent of a Thermal Unit, and M. Regnault's Observations on Steam. *Transactions of the R. S. of Edinburgh*, *20*, 261–298.

Young, T. (1807). *Thomas Young's Lectures on Natural Philosophy and the Mathematical Arts* (Vol. I). Thoemmes (rep. 2002).

Appendix B
Vis Viva: Leibniz, Mayer and Helmholtz

Mayer presents a quantitative relationship between falling force and force of motion. For this, he uses a relationship between falling height and velocity that came from the seventeenth century. In the study of falling graves, Galileo had arrived at the following result. In the first unit of time, the body travels through a certain space. In the second unit of time, the body travels three times this amount of space. In the third unit of time, it covers five times, etc. (Table B.1).

The velocity of the body in the first unit of time is 1 because it is given by the distance covered (let us take equal to 1) divided by the time spent (one unit). At the end of the second unit of time, calculating in the same way, the total distance covered is $1 + 3 = 4$ and the time spent is 2. Therefore, the average velocity is equal to 2. At the end of the third time unit, the distance traveled is $1 + 3 + 5 = 9$. Therefore, the average velocity is $9/3 = 3$. Thus, it can be said that the height is proportional to the square of the average velocity. This relationship was used by Leibniz.

In 1686, Leibniz published the paper *Brevis demonstratio erroris memorabilis Cartesii et aliorum circa legem naturae* … (Leibniz, 1686). What was Descartes' error, according to Leibniz? In 1644, Descartes held that the quantity of motion in the universe is constant (Descartes PP II, § 36). This meant that if one body loses motion, another body or bodies gain motion. According to Descartes, two bodies have an equal quantity of motion if the products of body's mass and velocity are equal

Table B.1 Galileo's results on falling bodies (Galilei, 1965, Vol. X, p. 115; Vol. VIII, p. 210)

Time	Distance covered per time unit	Distance covered at the end of each time unit
1	1	1
2	3	4
3	5	9
4	7	16
…	…	…

R. Lopes Coelho, *What Is Energy?*, History of Physics,
https://doi.org/10.1007/978-3-031-51855-3

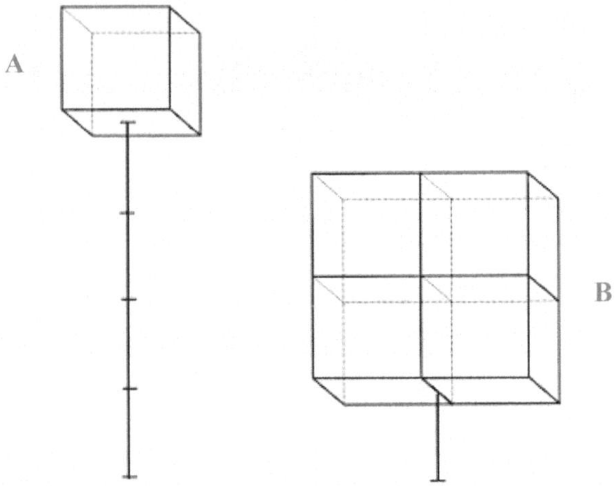

Fig. B.1 the force of body A ($m = 1, h = 4$) equals the force of B ($m = 4, h = 1$)

(Descartes, 1973, Part II, § 36). Thus, to agree with the universal law of conservation of motion, the quantity of motion before a (elastic) collision is equal to the quantity of movement after collision (ibid. §§ 46–52).[24] In 1669, Huygens shows that in elastic collisions, not only the quantity of motion but also the magnitude mv^2 is conserved.[25] In 1686, Leibniz argues that this magnitude is conserved in nature in general (Leibniz, 1971). His argument is as follows. A body with a mass of 4 units, at a height of 1 unit has as much force as a body of mass 1, at a height of 4 units (Fig. B.1), in short,

$$1_{mass} \times 4_{height} = 4_{mass} \times 1_{height}$$

This equality is not justified in the article. It is admitted. It is used afterwards to move to an equality where the velocities of the bodies appear. (This is what Leibniz was interested in to overthrow Descartes' law, which includes velocities). The average velocity of body A is equal to two units. The average velocity of body B is equal to one unit as seen above. Now, to maintain the admitted equality

$$m_A h_A = m_B h_B \tag{B.1}$$

we cannot use the product of mass and velocity, as established by Descartes' law, because this would give

$$1_A \times 2_A \neq 4_B \times 1_B$$

but rather, Leibniz claims,

[24] Impact was the most important phenomenon, because in the Cartesian universe there is no empty space. Therefore, there was no movement without impact.

[25] "La somme des produits faits de la grandeur de chaque corps dur, multiplié par le quarré de sa vîtesse, est toûjours la mesme devant & apres leur reencontre". (Huygens, 1669, p. 180)

$$1_A \times 2_A^2 = 4_B \times 1_B^2 \tag{B.2}$$

Hence, he argues that the magnitude mv^2 is conserved.[26]

Mayer neither used this equation nor Eq. B.1. Instead of this, he did the following. Taking weight and height as the cause of the fall and admitting that cause = effect, he justifies the equality between mass times height (cause) and mass times the square of velocity (effect). It follows then for each of Leibniz's bodies that

$$m_A h_A = m_A v_A^2 \tag{B.3}$$

$$m_B h_B = m_B v_B^2 \tag{B.4}$$

Therefore,

$$m_a v_A^2 = m_B v_B^2 \tag{B.5}$$

since the falling force is equal in Leibniz's bodies (B.1). As this was Leibniz's conclusion and this is justifiable by means of Mayer's claim—the quantity of force is indestructible, he argues that the conservation of living force is subsumed by his theory of force.

Vis viva is one of Helmholtz's ultimate forces. He uses, however, an algorithm according to which weight times distance gives ½ mv^2. For this reason, he changes the meaning of living force: living force becomes ½ mv^2 in his paper. There is, however, a difference between this velocity v and Mayer's. Lebniz's living force, which is the one that Mayer uses, is expressed by mv^2, but this velocity is the average velocity over the path. This is not the velocity that Helmholtz uses. As we have seen, Leibniz calculated the velocity from all the distance traveled by the body and all the time spent. To determine the distance traveled, he took the distance traveled in the first unit of time as a unit, as Galileo had done. Later, the distance traveled was given by ½ gt^2. If we calculate the average velocity as Leibniz did, we get ½ gt. Then the square of this, which is what Leibniz uses, is ¼ $(gt)^2$.

Another expression of the velocity is given by the derivation of $h = ½\, gt^2$. From here comes $v = gt$. This is the velocity of the body at the end of time t, not the average velocity of the body. Comparing the two, we get ½ $v^2 = 2{v^2}_{ave}$. In short, Helmholtz does not halve Leibniz's living force, because the velocity he uses is not Leibniz's.[27]

Some historians have argued that Mayer got it wrong, because he used mv^2 instead of ½ mv^2.[28] This criticism stems from the fact that both velocities are represented by the same letter. In fact, Mayer is referring to Leibniz. What he used was, therefore,

---

[26] He called this magnitude 'living force' (vis viva) later on (Iltis, 1971, p. 25).

[27] The confusion between the two velocities already existed in Coriolis (1829). He used the same algorithm as Helmholtz and also claimed that ½ mv^2 should be called living force. Both Coriolis and Helmholtz interpreted that the living force was halved.

[28] The Germann editors of Mayer's texts Weyrauch (Mayer, 1893) and Oettingen (Mayer, 1911) added the "corrected" expressions to the original equations.

that average velocity, which differs from the velocity that appears in the kinetic energy expression.[29]

References

Caneva, K. L. (2021). *Helmholtz and the Conservation of Energy: Contexts of Creation and Reception.* MIT Press.

Coriolis, G. (1829) *Du Calcul de l'Effect des Machines.* Paris: Carilian-Goeury.

Descartes, R. (1973). *Principia Philosophiae* (P. T. Ch. Adam, Ed. Vol. VIII-1). J. Vrin.

Galilei, G. (1965). *Le Opere di Galileo Galilei* (Vol. VIII, X). Firenze, G. Barbèra.

Huygens, C. (1669, 1929). Regles du mouvement dans la rencontre des Corps. In J. A. Vollgraff (Ed.), *Christiaan Huygens, Oeuvres complètes* (Vol. XVI, pp. 179–181). Martinus Nijhoff.

Iltis, C. (1971). Leibniz and the Vis Viva Controversy. *Isis*, *62*, 21–35.

Katzir, S. (2023) The use of the conservation of living force before Helmholtz, *Annals of Science* 80, 337–356. DOI: https://doi.org/10.1080/00033790.2023.2205429.

Leibniz, G. W. (1686). Brevis Demonstratio erroris memorabilis Cartesii, et aliorum circa legem naturae, secundum quam volunt a Deo eandem semper quantitatem motus conservari, qua et in re mechanica abutuntur. *Acta Eruditorum*, 161–163. In Leibniz, G. W. (1971). Mathematische Schriften, Vol. VI, C. I. Gerhardt (Ed.), Hildesheim: G. Olms Verlag.

Mayer, J. R. (1911). *Die Mechanik der Wärme: Zwei Abhandlungen (1842/1845)* (A. v. Oettingen, Ed. Vol. 180). Engelmann.

Mayer, J. R. (1893) *Die Mechanik der Wärme in gesammelten Schriften v. Robert Mayer.* J. Weyrauch (ed.). Stuttgart: Cotta.

[29] Considering the use of the conservation of living force in Mechanics before Helmholtz, Katzir (2023) criticized some of Caneva's theses on Helmholtz (Caneva, 2021). This discussion does not concern, however, the point presented above: Mayer's velocity v in mv^2 is not the same as v in Helmholtz's $\frac{1}{2}mv^2$.

Appendix C
Mayer and the Color of Venous Blood

On a boat trip to Indonesia, Mayer performed phlebotomies on all the crew members and was amazed at the bright color of the venous blood compared to the color of the blood he knew as a European doctor. He then concluded that there was a relationship between blood color and ambient temperature (Sect. 2.1).

Around the time of Mayer birth centennial celebration, Jentsch (1914) published a book in which the history of Mayer's discovery is addressed and an article in the German journal *Naturwissenschaften*, in which he reviewed other papers of the time on the color of venous blood. He concluded that there was no proof that venous blood in the equatorial zone was brighter (Jentsch, 1916). In 1954, Farber published an article on blood color to address Mayer's observation considering articles on venous blood color before Mayer's time. He stresses an experimental finding by Crawford who had placed a dog in water at 44.44 °C and found that after half an hour the venous blood acquired the color of the arterial blood (Farber, 1954, p. 6). This finding may lead one to think that the tropical environment of Java would have had a similar effect, which contradicts Jentsch's remark that there is no proof that the venous blood is brighter in the equatorial zone. Let us consider Crawford's experiment.

A dog with a temperature of 38.89 °C was immersed in water at a temperature of 45.55 °C. After half an hour, the dog's temperature was 42.78 °C and the water temperature was 44.44 °C. At this point, blood was drawn from an artery and a vein. "The venous blood assumed very nearly the hue of the arterial, and resembled it so much in appearance, that it was difficult to distinguish between them" (Crawford, 1788, p. 309). (This contrast with the initial conditions of the experiment: "the colour of the venous blood is a dark modena red; and that of the arterial a light scarlet." (ibid. p. 309)

The result of this experiment corroborates Mayer's observations, in that when the dog's environment became warmer, the colors of the arterial and venous blood were not easily distinguishable. There are, however, significant differences between the conditions of this experiment and those in which Mayer's observation took place. In Crawford's experiment, the dog's temperature increased by 3.89 °C. If we take the dog's initial temperature as its normal temperature and transfer this temperature

R. Lopes Coelho, *What Is Energy?*, History of Physics,
https://doi.org/10.1007/978-3-031-51855-3

difference to the human being, this will mean an increase in body temperature from 37 °C (considered the normal temperature) to around 40.89 °C. However, Mayer did not mention such a temperature increase. In addition, Crawford notes that the dog's respiration became "very rapid" after 11 min and still more rapid after 13 (ibid. p. 308), while Mayer did not mention any change in the speed of breathing.

In sum, although Crawford treated experimentally the variation in the color of venous blood due to the heat to which a body is subjected, the conditions of the experiment are significantly different from the conditions under which Mayer made his observation.

More recently, the doctor Rocha-Homem addressed Mayer's clinical observation. She studied the color of venous blood and concluded that it may be brighter in some circumstances for metabolic reasons (Rocha-Homem, 2015). This could explain Mayer's observation in a way that neither contradicts Jentsch's research nor interferes with Crawford's results.

References

Crawford, A. (1788) *Experiments and Observations on Animal Heat, and the Inflammation of Combustible Bodies; Being an Attempt of Resolve these Phenomena into a General Law of Nature*. 2nd Ed. J. Johnson.

Farber, E. (1954). The Color of Venous Blood. *Isis*, *45*, 3–9.

Jentsch, E. (1914). *Julius Robert Mayer seine Krankheitsgeschichte und die Geschichte seiner Entdeckung*. J. Springer.

Jentsch, E. (1916). Zur Geschichte der Entdeckung Julius Robert Mayer. *Die Naturwissenschaften*, *4*, 90–93.

Rocha-Homem, T. (2015). Robert Mayer: Conservation of Energy and Venous Blood Colour. *Advances in Historical Studies*, *4*, 309–313.

Appendix D
Imponderability: A Property of Mayer's Force

In addition to the two properties of force—indestructibility and transformability—called essential, (Sect. 2.1.1) there is a third—imponderability, whose justification is of a different kind. In nature, there are two sets of causes, matter and force, says Mayer. Matter is characterized by ponderability and impenetrability and force, by the opposite.[30] In the author's own synthesis, forces are quantitatively indestructible, qualitatively transformable and imponderable.[31]

'Imponderability' occupied a prominent place in the science of that time. Some authors divided their works into two parts: the first being concerned with ponderable matter, which roughly corresponds to Mechanics, and the second with imponderables, which included heat, light, electricity and magnetism.[32] In *Fundamentals of Physics and Chemistry*, Suckow (1813) entitled the second part of the book "On the imponderable fundamental substances of bodies". Muncke organizes his book *Theory of Nature*, 1829, in the same way. The second part of this book is entitled "Imponderable powers".

Suckow's 'fundamental substances' or Muncke's 'powers' were said to be imponderable because experiments had shown that unlike other matter, they were weightless. It had been, for instance, verified experimentally that a hot body had no more

[30] "Zwei Abtheilungen von Ursachen finden sich in der Natur vor, zwischen denen erfahrungsmäßig keine Uebergange stattfinden. Die eine Abtheilung bilden die Ursachen, denen die Eigenschaft der Ponderabilität und Impenetrabilität zukommt, Materien; die andere die Ursachen, denen letztere Eigenschaften fehlen, - Kräfte." (Mayer, 1842, p. 234)

[31] "Kräfte sind also: *unzerstörliche*, *wandelbare*, *imponderable* Objecte" (ibid. p. 234). The introduction in the text of the adverbs quantitatively and qualitatively is justified by the following passage: "Ursachen sind (quantitativ) *unzerstörlich* und (qualitativ) *wandelbare* Objecte" (ibid. p. 234).

[32] "*Unwägbare Stoffe*; *Imponderabilia*. Diejenigen Potenzen, welche die Erscheinungen des Lichtes, der Wärme, der Elektricität und des Magnetismus hervorbringen, werden unwägbare Stoffe, Imponderabilien genannt, weil sie sich von den übrigen bekannten materiellen Substanzen dadurch unterscheiden, daß sie nicht gewogen werden können". (Gehler, Vol. 5.2, p. 765)

R. Lopes Coelho, *What Is Energy?*, History of Physics,
https://doi.org/10.1007/978-3-031-51855-3

weight than the same cold body.[33] The imponderables were distinguished from the rest of matter by another property: they could not be enclosed permanently in a container. Hence, they were also called non-coercive.[34]

According to Muncke (1830), in a review article, there was a tendency to consider powers as substances, but there was no basis for affirming or denying this thesis.[35] The designation "imponderables", he adds, does not explain the nature of heat, light, electricity or magnetism and it is even unclear whether the designation is appropriate for them.[36]

References

Mayer, J. R. (1842). Bemerkungen über die Kräfte der unbelebten Natur. *Annalen der Chemie und Pharmacie, 42*, 233–240.

Muncke, G. W. (1829). *Handbuch der Naturlehre* (Vol. I). Universitäts-Buchhandlung C. Winter.

Gehler, J. S. T. (Ed.) (1830). *Gehler's Physikalisches Wörterbuch* (Vol. 5,2). Leipzig: Schwickert.

Suckow, G. A. (1813). *Anfangsgründe der Physik und Chemie nach den neuesten Entdeckungen.* Stage.

33 "Letzterer [Rumford] brachte nämlich Flaschen mit Weingeist, Quecksilber und Wasser auf empfindlichen Waagen ins Gleichgewicht, ließ sie von 61° F. bis 29° F. erkalten, bei welcher Temperatur das Wasser in Eis verwandelt war, und fand nicht den mindesten Unterschied bei der Anwendung einer Waage, welcher auf ein Millionth. des Totalgewichts noch einen Ausschlag gab". (Muncke, 1829, p. 376)

34 "Die genannten vier Potenzen heißen Incoercibilien, weil sie sich nicht auf gleiche Weise, als die wägbaren Stoffe, in Hüllen einschließen lassen". (Muncke, 1829, p. 369)

35 "[…] es belohnt sich daher kaum der Mühe, weitläuftige Untersuchungen darüber anzustellen, denn es läßt sich selbst nicht einmal ein Schluß für oder wider die Materialität derselben darauf gründen. Letzteres ist zwar verschiedentlich gesehen, aber sehr mit Unrecht". (Gehler, Vol. 5.2, p. 766)

36 "Aus allen diesen Betrachtungen folgt, daß man zwar die vier bekannten und ihnen ähnliche Potenzen mit dem Namen der Imponderabilien belegen könne, daß es dabei aber immer noch fraglich bleibt, ob dieser ihnen überhaupt oder in ganzer Strenge zukommt, wobei auf allen Fall das Wesen derselben keineswegs dadurch erklärt wird". (ibid. p. 770)

Appendix E
The Electrophorus

Electrophorus etymologically means 'carrier of electricity'. The instrument consists of a base, where the so-called resin cake is located, an upper plate and a handle, which serves to lift the upper plate (Fig. E.1). The resin cake is made of a mixture of resin, turpentine and beeswax. The plate is made of a metallic substance and the handle of a non-conductive material. Instead of a resin cake a glass disc could be used.

Here is a description of the instrument's operation from the time. By rubbing the resin cake (or a glass disk) and placing the plate on it base, one could observe the following.[37] If you touch it, you get a small spark. If you touch it with one

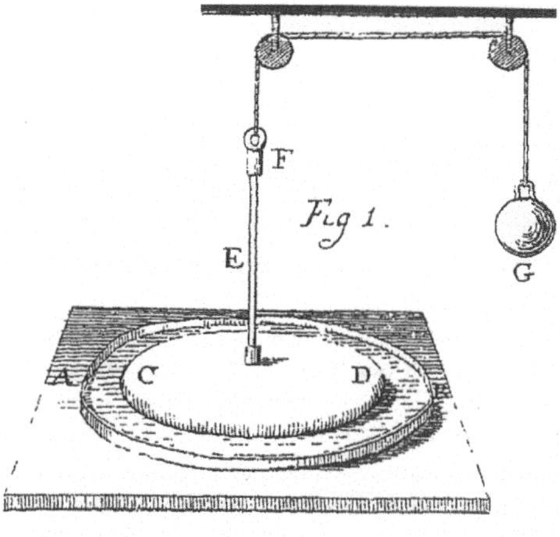

Fig. E.1 Wilcke's electrophorus, 1778: the resin cake (AB), upper plate (CD) and the handle (E). The pulleys are only for lifting the upper plate more easily

[37] "Man errege die E. des Kuchens durch Reiben. Ist derselbe, wie gewöhnlich, von einer harzigen Composition bereitet, so wird die Erregung am besten gelingen, wenn [...] Bedient man sich einer Glasscheibe, so dient zur Reibung derselben am besten [...]". (Gehler, 1827, Vol. 3.2, p. 736–7)

R. Lopes Coelho, *What Is Energy?*, History of Physics,
https://doi.org/10.1007/978-3-031-51855-3

finger and touch the cake pan with another, then you get a considerable shock.[38] After touching it with your fingers, neither the plate nor the cake pan shows any sign of electricity.[39] If you raise the plate to a sufficient height and touch it again, you get one or more sparks, which will be stronger if you simultaneously touch the plate and cake pan.[40] The process could be repeated many times. A single rubbing is sufficient, under favorable conditions of air humidity, to keep the instrument in action for months. This is why it was called 'perpetual carrier of electricity'.[41] The designation *perpetual electrophorus* was suggested by Volta, who is credited with the discovery of the instrument in 1775.[42]

Wilke had made known an analogous instrument, in 1762, with the difference that in place of the resin cake appeared a glass plate (Wilke, 1762). In 1778, he wrote an article on Volta's electrophorus, which he explained by means of the theory he had developed for his own apparatus (Wilke, 1778).

Volta's apparatus became especially important because it allowed the visualization of positive and negative electricity. This was achieved by Lichtenberg. He found that the resin powder, if dropped on the cake, acquired shapes, which were essentially radial or circular depending on the type of electricity, positive or negative (Fig. E.2).

The electrophorus was an irreplaceable instrument in a laboratory of the time.[43] For the explanation of the observed phenomena, there were several theories, as reported in the review article in Gehler's *Physical Dictionary* (Gehler, 1827). Mayer does not commit himself to any of the theories. In fact, he does not need the explanations either, because the way he uses the instrument in his theory depends only on observable elements: the movement of the upper plate and the electrical effects referred to above (Sect. 2.1.2).

---

[38] "Setzt man den Deckel [...] auf die Basis, welche nicht isolirt ist, und berührt ihn, so erhält man einen kleinen schneidenden Funken, berührt man dagegen mit dem einen Finger die Form, mit dem andern den Deckel, so fühlt man einen erschütternden Schlag". (ibid. p. 737)

[39] "Nach diesen Berührungen zeigt weder der Deckel noch die Form weiter einige Spuren von E." (ibid. p. 737)

[40] "Hebt man hierauf den Deckel mit den Schnüren oder Glasstange isolirt auf, entfernt ihn genugsam von der Basis, und berührt ihn wieder, so erhält man einen oder mehrere stechende Funken [...] Diese Funken sind stärker, wenn man den Deckel und die Form zugleich, als wenn man nur den Deckel allein berührt hat". (ibid. p. 737)

[41] "Das [...] Verfahren läßt sich, so oft man will, wiederholen, ohne daß der Kuchen etwas merkliches von seiner E. verliert, bis ihm endlich nach längerer Zeit Luft und Feuchtigkeit dieselbe entziehen. So kann man von einer einzigen Reibung oft Monate lang el. Funken erhalten, daher das Instrument der *beständige Elektricitätsträger* genannt worden ist". (ibid. p. 738)

[42] "E come mercè un tal rifondere quella elettricità che la mia macchina conserva già da sè assai lungo tempo, si può rendere non che durevole per lunghissimo tratto, ma perenne, così al nome di *Elettroforo*, il più acconcio per una tal machina, ho aggiunto il predicato di *perpetuo* che a tutto buon diritto gli conviene". (Volta, 1918, p. 178)

[43] See Schmidt (1813, pp. 443–447), Suckow (1813, p. 474 ff), Biot (1816, pp. 374–82), J. T. Mayer (1820, pp. 464–473), Muncke (1829, pp. 749–755).

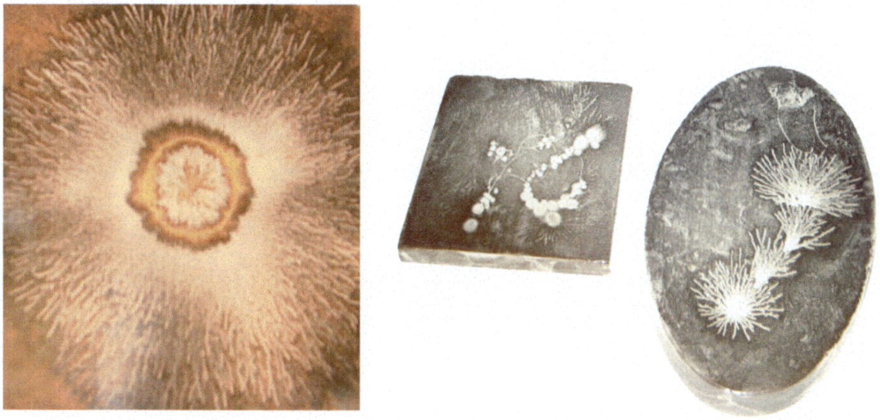

a b

Fig. E.2 Lichtenberg's images in the laboratory: **a** is similar to the image published by Lichtenberg (1806, T. II); **b** represents negative and positive electricity

References

Biot, J. B. (1816). *Traité de physique et mathématique* (Vol. II). Deterville.

Gehler, J. S. T. (1827). Erscheinungen und Gebrauch des Elektrophors. In *Physikalisches Wörterbuch* (Vol. 3,2, pp. 736–742). Leipzig: Schwickert.

Lichtenberg, G. C. (1806). Von einer neuen Art die Natur und Bewegung der elektrischen Materie zu erforschen. Erste Abhandlung. In L. C. Lichtenberg, F. Kries (Eds.), *Georg Christoph Lichtenberg's Vermischte Schriften* (Vol. 9). Heinrich Dieterich.

Mayer, J. T. (1820). *Anfangsgründe der Naturlehre* (4 ed.). Dietrich.

Muncke, G. W. (1829). *Handbuch der Naturlehre* (Vol. I). Universitäts-Buchhandlung C. Winter.

Schmidt, G. G. (1813). *Handbuch der Naturlehre* (2 ed.). Giessen.

Suckow, G. A. (1813). *Anfangsgründe der Physik und Chemie nach den neuesten Entdeckungen.* Augsburg: Stage.

Volta, A. (1918). *Le Opere di Alessandro Volta* (Vol. III). Ulrico Hoepli.

Wilcke, J. C. (1762). Von den entgegesetzten Eleckticitäten. *Neue Abhandlungen der königlichen Schwedischen Akademie der Wissenschaften 24*, 213–235.

Wilcke, J. C. (1778). Untersuchungen der bey Herrn Voltas neuen Electrophoro perpetuo vorkommenden elektrischen Erscheinungen. *Neue Abhandlungen der königlichen Schwedischen Akademie der Wissenschaften 39*, 54–78.

Appendix F
Mayer and Holtzmann

Holtzmann published a book in 1845—*On the heat elasticity of gases and vapors*—in which he presents a value for the relationship between heat and work. This value is close to the mechanical equivalent of heat determined by Mayer. Let us compare the algorithms of both and the conceptual approach to the phenomena in question.

According to Holtzmann, heat is measured by its effects. These can be temperature rises, mechanical work, or both (Holtzmann, 1845, p. 7). If heat is measured through its effects and work is one of them, then heat can be measured by means of mechanical work. If heat is measured through work, then it makes sense to determine the unit of heat in terms of work. This is what Holtzmann is going to do. The unit of heat that he symbolizes by '*a*' is given by the ratio between the work done by the gas and the heat given to the gas, (ibid. p. 8)

$$a = \frac{pressure \times variation\,of\,volume}{heat\,given\,to\,the\,gas}$$

In determining the mechanical equivalent of heat, Mayer had also broken down the effect of heat into two: temperature rise and work. In this case, work was given by the product of weight and height (Sects. 2.1.1 and 2.2.2). The mechanical equivalent of heat (MEH) was given as

$$MEH = \frac{weight \times height}{heat\,given\,to\,the\,gas}$$

The gas in question in both authors is atmospheric air. The heat given to the gas is expressed in both authors by the difference between specific heat at constant pressure and constant volume. In the calculation, both use the ratio between these heats $\frac{C_p}{C_v}$. However, this ratio is equal to 1.415 according to Holtzmann (ibid. p. 12) and equal to 1.421 according to Mayer (1845, p. 15). Holtzmann's 'volume variation' and Mayer's 'height' do not differ significantly because the volume change is due to the expansion of the gas. Now, if the gas expands, it pushes the piston to a certain height.

R. Lopes Coelho, *What Is Energy?*, History of Physics,
https://doi.org/10.1007/978-3-031-51855-3

Thus, we understand that the coefficient of expansion of the gas (0.003668) used by Holtzmann (ibid. p. 13) is close to the 'height' indicated by Mayer (0.00365). If we use Holtzmann's algorithm but enter the Mayer values mentioned before, Holtzmann's unit of heat becomes equal to 367 kg m.[44] This is the value of the mechanical equivalent of heat determined by Mayer in 1845.

Conclusion. Holtzmann's unit of heat conforms to Mayer's mechanical equivalent of heat. The difference between the authors is in the approach. In one case, what heat does in terms of work is considered. Here the agent is heat, which remains in the process. In the other, the heat disappears to give rise to work—heat is transformed into motion. With Holtzmann, heat can be a substance. With Mayer, it is not a substance but a form of force. In short, the conceptions are different, the determined value is the same.[45]

References

Holtzmann, C. (1845). *Über die Wärme Elastizität der Gase und Dämpfe*. T. Loeffler.

Holtzmann, C. (1866). *Mechanische Wärme-Theorie*. Metzler.

Mayer, J. R. (1845). *Die organische Bewegung in ihrem Zusammenhange mit dem Stoffwechsel. Ein Beitrag zur Naturkunde*. Drechsler.

[44] The value obtained by Holtzmann was 374 kg m (ibid. p. 13).

[45] Holtzmann (1866) gives priority to Mayer and admits he made a mistake.

Appendix G
Mohr and Mayer

In 1837, Mohr wrote an article—*On the nature of heat*—in which he defends the thesis that, besides the chemical elements, there is an agent in nature. This is the force.

The impetus for this article came from one by Melloni (1836), in which he shows the polarization of heat and explores the analogy between this and the polarization of light (Mohr, 1837, p. 419). With this idea of heat being analogous to light (ibid. p. 441), Mohr explains thermal phenomena by means of the concepts of frequency and amplitude. He connects frequency to the temperature detectable by the thermometer: higher frequency means higher temperature (ibid. p. 427). If a given substance receives heat and the temperature does not increase, the increase in amplitude with maintenance of frequency justifies the heat received. The cooling of a gas in a rarefaction is explained by the increase of amplitude at the expense of frequency (ibid. p. 437).

So far, heat is motion. Heat will also become force. Let us see how this is done. Mohr argues that what reacts to a force is also a force. Now, heat overcomes the cohesive force of the particles. Therefore, heat is a force.[46]

It turns out that heat has other effects than mechanical. Seebeck (1822–23) had shown that heating the junction of two metals generates electric current (ibid. p. 443). Mohr looked at this phenomenon in the following way: heat appears as electricity. Since he then admits that heat gives electricity, he infers that there must be something common to both heat and electricity. Now, heat is force, as seen above. Therefore, electricity will also be force. Going through several other phenomena in this way, he comes to the conclusion that there must be a single force in nature that accounts for what all these phenomena have in common. This leads to the thesis mentioned at the beginning of this section:

[46] "Die Wärme erscheint in unzähligen Fällen als eine Kraft. [...] Die Wärme hebt ebenfalls die Kohäsion der Körper auf, was aber eine Kraft aufhebt, muss selbst eine Kraft seyn". (ibid. p. 421)

R. Lopes Coelho, *What Is Energy?*, History of Physics,
https://doi.org/10.1007/978-3-031-51855-3

> "Besides the known 54 chemical elements, there is only one agent in the nature of things, and that is force; it can appear [..] as motion, chemical affinity, cohesion, electricity, light, heat, and magnetism". (ibid. p. 442)[47]

According to Mayer, there is a force that pervades the organic and the inorganic. This force has five forms (Sect. 2.1.2). Some of these coincide with Mohr's forces. If we take the concepts out of context, the theses seem similar. There are, however, major differences. Mayer starts from the equation '*causa aequat effectum*'. 'Force' is the name he gives to cause and effect. The constancy of force in nature, which is the core of his axiom, (Sect. 2.1.2.1) comes from that initial equation. Based on it, Mayer determines the mechanical equivalent of heat. A comparison between both Mayer and Mohr is made by Mohr himself.

In 1868, when the energy theory was already important, Mohr wrote about his 1837 paper. He does not claim for himself the credit for the discovery of energy, but he quotes several passages from his article that are in line with what was said about energy (Mohr, 1868, pp. 39–45; see also Mohr, 1869, pp. 82–4).

References

Melloni, M. (1836). Mémoire sur la Polarisation de la Chaleur. *Annales de chimie et de physique, 61*, 375–410.

Mohr, F. (1837). Ueber die Natur der Wärme. *Zeitschrift für Physik und verwandten Wissenschaften, V*, 419–445.

Mohr, F. (1868). *Mechanische Theorie der chemischen Affinität und die neuere Chemie*. F. Vieweg und Sohn.

Mohr, F. (1869). *Allgemeine Theorie der Bewegung und Kraft, als Grundlage der Physik und Chemie*. F. Vieweg und Sohn.

Seebeck, T. (1822–23). Magnetische Polarisation der Metalle und Erze durch Temperatur-Differenz. *Abhandlungen der Königlichen Akademie der Wissenschaften zu Berlin*, 265–373.

[47] "Ausser den bekannten 54 chemischen Elementen gibt es in der Natur der Dinge nur noch ein Agens, und dieses heisst Kraft; es kann unter den passenden Verhältnissen als Bewegung, chemische Affinität, Cohäsion, Elektricität, Licht, Wärme und Magnetismus hervortreten". (Mohr, 1837, p. 442)

Appendix H
The Magneto-Electricity

The term 'magneto-electricity' has fallen into disuse. The phenomena called magneto-electric are today call 'electromagnetic'. At that time, electromagnetism and magneto-electricity had opposite meanings.

In 1820, Oersted showed that the passage of electric current had an effect on a magnetic needle (Oersted, 1920, p. 214). Because electric current comes first and then the effect on the magnetic needle, Oersted called this phenomenon *electromagnetic*. A few years later, Seebeck showed that heating the junction of two metals gave rise to electric current (Seebeck, 1822–23, p. 265). Oersted proposed then to call this phenomenon *thermoelectric*.[48] In 1831, Faraday verified that the movement of a magnet in the vicinity of a copper coil gave rise to electric current. Since in this case we have first the movement of the magnet and then electric current, the phenomenon was called *magneto-electric*.[49] Hence, Joule's machine was called magneto-electric machine and the electric current thus produced, magneto-electricity.

This way of designating this kind of phenomena, which was common at the time, gave an account of what was being observed. A pair of terms refers to what is observed and requires no further explanation. Can we say that a pair of terms refers to a conversion, for example, from heat into electricity? A necessary condition for this is that heat exists, and electricity exists. The weak point concerns electricity. At the time, physicists were faced with the question of whether "electricities" derived from different sources had identical nature (Faraday, 1839, p. 102). Indeed, the thermo-electric phenomenon created a new kind of electricity. The same holds for the magneto-electric phenomenon. In this context, it had been meaningless to talk about *conversion* of:

---

[48] "Il faudra désormais distinguer cette nouvelle classe de circuits électriques par une dénomination significative; et comme telle je propose l'expression de circuits thermo-électriques ou peut-être thermélectriques". (Oersted, 1823, p. 199)

[49] "The various experiments of this section prove, I think, most completely the production of electricity from ordinary magnetism" (Faraday 1832, p. 138). "I propose to call the agency thus exerted by ordinary magnets, *magneto-electric* or *magnelectric* induction" (ibid. p. 139).

R. Lopes Coelho, *What Is Energy?*, History of Physics,
https://doi.org/10.1007/978-3-031-51855-3

- heat into electricity, because, if it were conversion, it should have been conversion from heat into thermo-electricity;
- magnetism into electricity, because it should have been conversion from magnetism into magneto-electricity;
- electricity into magnetic effect, because it should have been conversion from voltaic electricity into magnetism.

In sum, the pair of terms that characterized these kinds of phenomena (electro-magnetic, magnetic-electric, thermo-electric) had no connection with 'conversion processes' at that time.[50] It is true that Mayer integrated these and other types of phenomena as 'transformations'. 'Transformation', however, does not have the sense of conversion. These terms differ from each other from the ontological point of view (Sect. 2.2.5).

References

Faraday, M. (1832). Experimental Researches in Electricity. *Philosophical Transactions of the Royal Society of London, 122*, 125–162.
Faraday, M. (1839). *Experimental Researches in Electricity* (Vol. 1). Taylor.
Heimann, H. (1973). Conversion of Forces and the conservation of Energy. *Centaurus, 18*, 147–161.
Oersted, H. C. (1823). Nouvelles Expériences de M. Seebeck sur les actions électro-magnétiques. *Annales de Chimie et de Physique 22, 199–201, 22*, 199–201.
Oersted, H. C. (1920). Experimenta circa effectum conflictus electrici in acum magneticam. In T. R. D. S. o. Sciences (Ed.), *H. C. Oersted: Scientific Papers* (Vol. II). Andr. Fred. Høst and Son.
Seebeck, T. (1822–23). Magnetische Polarisation der Metalle und Erze durch Temperatur-Differenz. *Abhandlungen der Königlichen Akademie der Wissenschaften zu Berlin*, 265–373.

[50] See also Heimann (1973, pp. 147–8).

Appendix I
Heat Is Motion: Rumford and Joule

Joule refers to Rumford in the post scriptum of the 1843 article (Joule, 1884, p. 157) and again in the 1850 paper, for having defended the thesis that heat is motion (Joule, 1850, p. 61). Indeed, the idea of heat being motion had already existed in the eighteenth century and was based on experiments (Appendix A). Joule's thesis on the nature of heat was therefore not new.[51] Let us consider Rumford's experiment to judge on Joule's novelty.

Rumford's experiment consists of inserting a metal piece to be drilled and the drill into a container with water and measuring the temperature of the water from time to time (Rumford, 1798, p. 90). The drill performs 32 revolutions per minute. The water temperature increases continually (Fig. I.1).

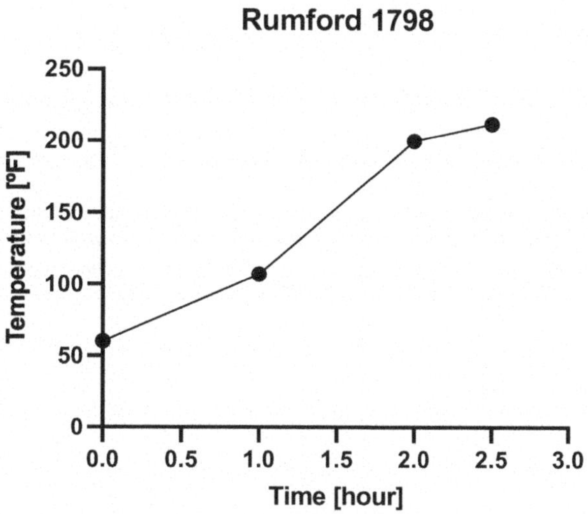

Fig. I.1 The experiment started with the water at 60° Fahrenheit. After one hour, the water temperature was 107°; after two hours, 200° and after two and a half hours, the water boiled (ibid. pp. 91–92)

[51] In the 1850 paper, Davy's experiment (Davy 1799) is also referred to (Joule, 1850, p. 62).

R. Lopes Coelho, *What Is Energy?*, History of Physics,
https://doi.org/10.1007/978-3-031-51855-3

Based on this experiment, the answer to the question in the article 'what is heat' was the following:

> "it appears to me to be extremely difficult, if not quite impossible, to form any distinct idea of any thing, capable of being *excited*, and *communicated*, in the manner the heat was excited, and communicated in these experiments, except it be MOTION". (ibid. p. 99)

Let us compare this to Joule's experiments. Generating heat through the magneto-electric machine, as Joule did, was new, just in the sense that Rumford did not even know the machine. However, magneto-electricity was just a means of producing heat through motion. Producing heat through motion was nothing new. The thesis—heat is motion—was not new either. There is, however, one new point. Joule not only talks about increasing heat by mechanical means, but also about decreasing heat. He wrote:

> "We have therefore in magneto-electricity an agent capable by simple mechanical means of destroying or generating heat". (Joule, 1884, p. 146)

All this appears in Part I of Joule's paper. In the second, Joule takes up some of the previous experiments with a new purpose: to verify whether there is a proportionality between the mechanical action involved and the heat produced (Joule, 1884, p. 149). This second part, entitled "On the mechanical value of heat", provides the novelty of the paper. This value was determined experimentally and made it possible to make predictions.

References

Davy, H. (1799). An Essay on Heat, Light, and Combinations of Light. In J. Davy (Ed.), *Collected Works* (Vol. 2, pp. 2–86). Smith, Elder and Co.

Joule, J. P. (1850). On the Mechanical Equivalent of Heat. *Philosophical Transactions of the R. S. of London* 140, 61–82.

Joule, J. P. (1884). On the Calorific Effects of Magneto-Electricity, and on the Mechanical Value of Heat. In T. P. Society (Ed.), *The Scientific Papers of James Prescott Joule* (Vol. I, pp. 123–159).

Rumford, B. C. (1798). An inquiry concerning the Source of the Heat which is excited by Friction. *Philosophical Transactions of the Royal Society of London, 88*, 80–102.

Appendix J
Gay-Lussac, Dulong and Joule

I recall that in 1845, Joule carried out gas experiments: condensation, rarefaction and rarefaction followed by condensation (Sect. 2.2.2). This type of experiment had already been carried out by Dulong and Gay-Lussac. The results obtained were in line with those of Joule. There is, however, a difference between these experiments and Joule's. It concerns heat by friction.

Dulong had conducted experiments in which the volume of a given amount of gas is suddenly reduced; and the reverse, the volume of a given amount of gas is suddenly increased. He presented the results of these experiments as follows:

> "It would thus be the case with compound gases as with simple gases, and we would be led to this general law remarkable for its simplicity, namely: 1. *that equal volumes of all elastic fluids taken at the same temperature and under the same pressure, being compressed or suddenly expanded by the same fraction of their volume, give off or absorb the same* ABSOLUTE QUANTITY OF HEAT; 2. *that the resulting variations in TEMPERATURE are inversely proportional to their specific heat at a* CONSTANT VOLUME".[52]

These experiments conform to the caloric theory. Heat exists in bodies, therefore also in gases. When a gas is pressed, heat comes out.[53] It is thus understandable that condensation gives rise to heat. If the gas has more space, it can receive heat. Therefore, it draws heat from its surroundings. Dulong shows that the heat absorbed in the latter case is equal to the heat released in condensation if the other items of the experiment are kept the same.

In the condensation and rarefaction experiments, Joule immerses the orifice through which the gas passes into water (Fig. J.1).

---

[52] "Il en serait donc des gaz composés comme des gaz simples, et nous serions conduits à cette loi générale remarquable par sa simplicité, savoir: 1°; *que des volumes égaux de tous les fluides élastiques pris à une même température et sous une même pression, étant comprimés ou dilatés subitement d'une même fraction de leur volume, dégagent ou absorbent la même* QUANTITÉ ABSOLUE DE CHALEUR; 2°; *que les variations de* TEMPÉRATURE *qui en résultent sont en raison inverse de leur chaleur spécifique à* VOLUME CONSTANT". (Dulong, 1829, pp. 155–6)

[53] Explaining the phenomenon through caloric theory would offer more information. In the present context, this information need not be considered.

R. Lopes Coelho, *What Is Energy?*, History of Physics,
https://doi.org/10.1007/978-3-031-51855-3

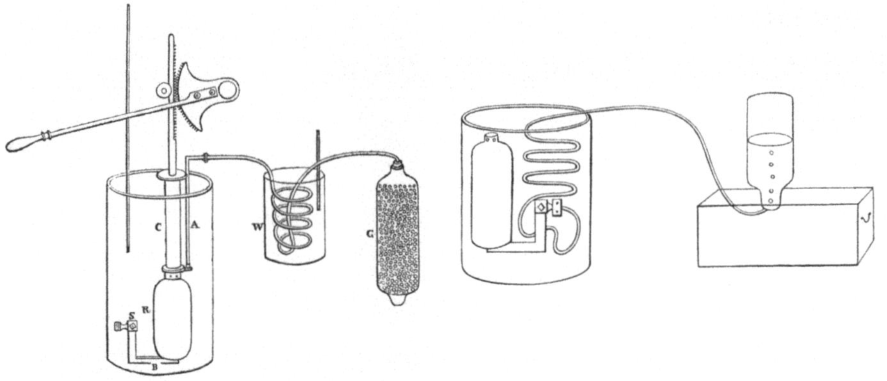

Fig. J.1 the configurations of the condensation and rarefaction experiments show the stopcock immersed in the recipients where the temperature of the water is measured

Therefore, the heat of friction will arise in both cases. In condensation, this heat is added to the proper heat of condensation, increasing the heat released. In rarefaction, the cooling is slightly counteracted by the heat of friction.[54]

In sum, Dulong's experiments are consistent with those of Joule, who in fact quotes him (Joule, 1845b, p. 187). The difference is that Joule includes the heat of friction, unlike Dulong. Thus, what is really in question is the heat of friction. Joule argues then that his concept of heat provides an explanation for the heat of friction, unlike the theory of heat as a substance.[55]

Let us move on to the rarefaction followed by condensation experiments. Joule performed two types of experiments (Sect. 2.2.2). In one case, the containers in which the gas is at considerable pressure and the exhausted container are immersed in different water-filled vessels. The stopcock is immersed in a third vessel (Fig. J.2). In the other case, the two containers and the stopcock area are immersed in a single vessel. The former case is similar to Gay-Lussac's experiments.

Gay-Lussac (1807) carried out experiments in which he determined the heat absorbed in rarefaction and released in condensation (Fig. J.3).

In these experiments, the pressures are much lower than in Joule's experiments: 1, 0.5 and 0.25 atm. If we take the value of the 'lost' heat in the rarefaction as negative and the 'gained' heat as positive, as Joule did, we have (Table J.1).

In Joule's experiments we have

$$-2.36^\circ + 2.38^\circ = 0.02^\circ.$$

[54] "These results are inexplicable if heat be a substance. If that were the case, the same quantity of heat would have been absorbed by the rarefaction [...] as was evolved by the corresponding condensation". (Joule, 1845b, pp. 185–6)

[55] Friction was the weakness of the theory (Rankine 1855, p. 212; Fox 1971, pp. 99–103).

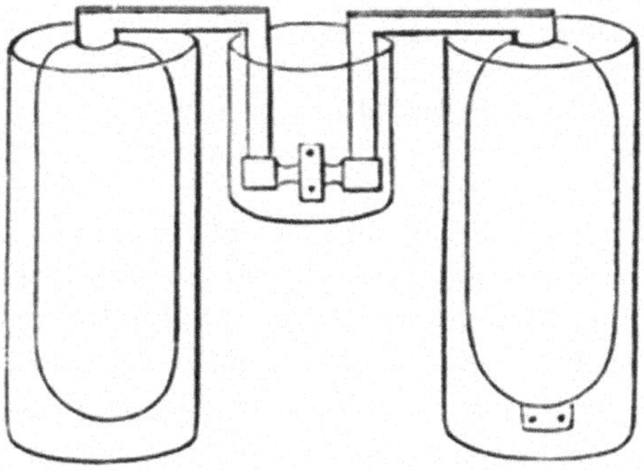

Fig. J.2 Schema of Joule's experiment (Joule, 1845b, p. 183)

Densité de l'air exprimée par le baromètre.	Froid produit dans le ballon n°. 1.	Chaleur produite dans le ballon n°. 2.
0^{m},76. . .	0°,61 . . .	0°,58
0^{m},38. . .	0°,34. . .	0°,34
0^{m},19. . .	0°,20. . .	0°,20

Fig. J.3 The table of Gay-Lussac's results (Gay-Lussac, 1807, p. 188)

Table J.1 Gay-Lussac's results

Cooling	Heating	Sum
– 0.61°	0.58°	– 0.03
– 0.34°	0.34°	0
– 0.20°	0.20°	0

Unlike Gay-Lussac, Joule also measures the temperature at the stopcock, 0.31 °F. If we add up the three values, we get then

$$-2.36^\circ + 2.38^\circ + 0.31^\circ = 0.33^\circ.$$

Joule uses these kinds of experiments to defend the correlation between heat and work: if there is no work, no variation of heat occurs. This correlation is demonstrated in the first type of experiments (gas containers and stopcock are immersed in the same water). However, based on the results obtained when containers and stopcock are taken separately (0.33 °F), a slight temperature rise would be expected when the

Table J.2 The relationship between work and heat according to Joule (1845b)

Δwork > 0	Δheat > 0
Δwork < 0	Δ heat < 0
Δwork = 0	Δheat = 0

containers and stopcock are immersed in the same water.[56] Therefore, the argument could be: with rarefaction followed by condensation, the temperature difference is close to zero. (This is justified by the values $-2.36 + 2.38$ as well as those obtained by Gay-Lussac.) The heat that occurs in the experiments with the objects immersed in the same water is due to heat by friction.

In sum, each one of the three Joule's experiments was not significant since the results were predictable. Taken together, however, they corroborate the correlation between heat and work (Table J.2).

Joule's argument focuses on the heat by friction. He points out the advantages of the concept of heat as motion by explaining this kind of heat. His series of experiments gain further relevance with the calculation of the mechanical equivalent of heat.[57]

References

Dulong, P. L. (1829). Recherches sur la chaleur spécifique des fluides élastiques. *Annales de chimie et de physique, 56*, 113–159.

Fox, R. (1971). *The Caloric Theory of Gases. From Lavoisier to Regnault.* Clarendon Press.

Gay-Lussac, J. L. (1807). Premier essai pour déterminer les variations de température qu'éprouvent les gaz en changeant de densité, et considérations sur leur capacité pour le calorique. *Mémoires de Physique et de Chimie de la Société d'Arcueil, I*, 180–203. (Maurice Crosland. New York, London: Johnson Reprint Corporation, 1967)

Joule, J. P. (1845b). On the Changes of Temperature produced by the Rarefaction and Condensation of Air. Philosophical. *Philosophical Magazine, XXVI*, 369–383.

Rankine, W. (1855). Outlines of the science of energetics. *Edinburgh New Philosophical, 2*, 120–141.

[56] A warming is recorded but assigned to the room temperature (Joule, 1845b, p. 182).

[57] This was Joule's goal in carrying out these experiments (Joule, 1845b, pp. 172, 180).

Appendix K
Berthollet and Hirn

Hirn and Berthollet have performed similar experiments about half a century apart. Berthollet (1809) interprets the results by means of the heat-substance theory. Hirn (1862) uses his experiment to determine the mechanical equivalent of heat. The point here is *sui generis*: If Hirn had performed Berthollet's experiments and not performed only a part of them, he would not have arrived at the mechanical equivalent of heat.

Let us start with Hirn's experiment. A heavy bar is impacted by a moving bar (Fig. K.1). Between the two bars there is a piece of lead, as illustrated in the figure. After impact, the temperature of the lead piece increases. Hirn determines the work done in crashing the piece of lead and the heat developed. Putting the two values into an equation, he obtains the mechanical equivalent of heat (Hirn, 1862, p. 111) [This became a new method of determining the mechanical equivalent of heat (Kipnis, 2014, 2028)].

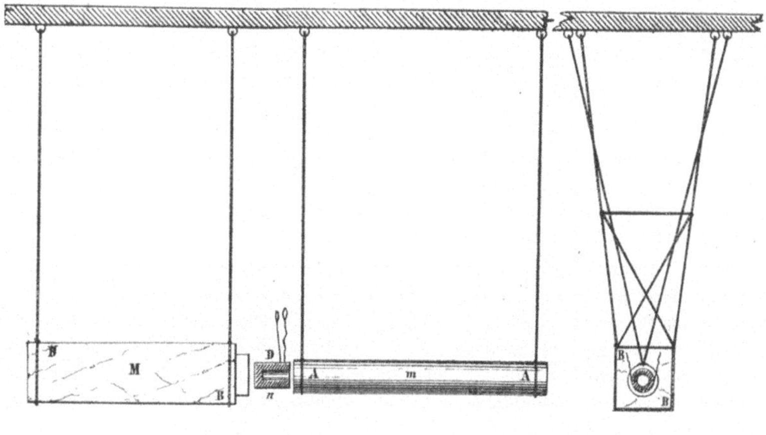

Fig. K.1 This is the Hirn experimental schema. The larger of the two pieces, on the lefthand side, is 941 kg and the smaller (m) is 350 kg. (Hirn 1862, p. 108)

R. Lopes Coelho, *What Is Energy?*, History of Physics,
https://doi.org/10.1007/978-3-031-51855-3

In 1809, Berthollet, together with Pictet and Biot, carried out metal compression experiments for which they used a "balancier".[58] Instead of lead, they used gold, silver, copper, iron and bronze blanks. After impact, the metal piece was placed in a small container with water and the water temperature measured. They verified that the temperature increased. So far, Berthollet, Pictet and Biot's result could be conform with Hirn's. Unlike Hirn, they continued the experiment in the following way. They let the bodies return to room temperature and performed the experiment again.[59] They put the metal piece in the water and measured the temperature. The water temperature increased, but less than in the first run of the experiment. They performed the experiment in the same manner a third time. The increase of the water temperature was lower than in the second run of the experiment. In the fourth run, the water temperature increase was even less or not detectable (Berthollet, 1809, p. 444). From the fourth on, there was no change in the water temperature. There was impact but no heat appeared. The authors came then to the conclusion that there is a relationship between the compression of the bodies and the heat developed; and that if the dimensions of the bodies cannot be further decreased, no more heat is produced, no matter how strong the mechanical action is.[60]

In sum, the work done is roughly the same in all previous experiments, but the amount of heat differs. If we were to use Hirn's algorithm with Berthollet's results, we would have the mechanical equivalent of the heat varying between zero and a maximum value—that of the first experiment.

One could argue that, in the second and subsequent runs of the experiment, we do not obtain the mechanical equivalent of heat because there is a certain factor that is not being taken into account. Admitting this argument, it follows that Hirn's algorithm is incomplete because such a factor is not provided.

Berthollet, Pictet and Biot could not have concluded that there is a relationship between work and heat. With the same quantity of work, they obtained different amounts of heat, including no heat. Under these circumstances, there was no reason to accept that heat is motion.

---

[58] "Il y a quelques années que, dans l'espérance de jeter du jour sur l'origine de la chaleur, qui provient de la compression et du frottement, je formai le projet d'examiner, à l'aide d'un balancier, les effets de la compression des métaux [...] je priai MM. Pictet et Biot de concourir avec moi à ces experiences". (Berthollet, 1809, p. 441)

[59] "Je fis préparer des flaons d'or, d'argent, de cuivre, de fer et de bronze, tous de même dimension, pour les soumettre à l'action du balancier [...] Pour déterminer la chaleur que les pièces de métal acquéroient par le choc du balancier [...] on préféra bientôt de jeter promptement la pièce dans une quantité d'eau suffisante pour la recouvrir [...] On soumettoit une pièce aux coups d'un balancier mis en mouvement par deux hommes forts et habitués à ce travail; on déterminoit la chaleur acquise; on laissoit reprendre à la pièce une température exactement semblable à celle du balancier; on lui faisoit subir un nouveau coup, et [...]". (ibid. pp. 442–3)

[60] "Il résulte de ce qui précède, que la chaleur qui est produite par le choc et la compression dans les corps qui n'éprouvent pas de changement chimique est uniquement due aux changemens de dimension qu'éprouvent ces corps, et lorsque les dimensions ne peuvent plus être diminuées, le choc quelque violent qu'il soit ne cause point de chaleur". (ibid. p. 447)

References

Berthollet, C. L. (1809). Notes sur divers objects. *Mémoires de Physique et de Chimie de la Société d'Arcueil*, *II*, 441–448. (New York: Johnson)

Hirn, G.-A. (1862). *Exposition analytique et expérimentale de la théorie mécanique de la chaleur*. Mallet-Bachelier.

Kipnis, N. (2014). Thermodynamics and Mechanical Equivalent of Heat. *Sci & Educ*, *23*, 2007–2044.

Appendix L
Newton and the Conservation of Energy

Peter Tait wrote:

> "so far as experimental facts were known in Newton's time, he had the Conservation of Energy complete; the cases of apparent loss by impact, friction, & c. were not then understood". (Tait, 1865, p. 57)[61]

Let us consider the interpretation at that time of the 'apparent loss by impact' case:

> "Two bodies void of elasticity, meeting together with equal contrary forces, both lose their motion". (Clarke's 4th fourth Reply, § 38, Alexander, 1956, p. 52)

This passage appears in the Leibniz-Clarke Correspondence (1715–16), where the latter represented Newton (Koyré & Cohen, 1962, pp. 64–7). Clarke argues that the bodies in that collision come to rest and need a new force to move. Therefore, he concludes, the amount of force has been lost.

Leibniz answered:

> "The author objects, that two soft or un-elastic bodies meeting together, lose some of their force. I answer, no. 'Tis true, that their wholes lose it with respect to their total motion; but their parts receive it, being shaken [internally] by the force of the concourse. And therefore that loss of force, is only in appearance. The forces are not destroyed but scattered among the small parts. The bodies do not lose their forces; but the case here is the same, as when men change great money into small". (Leibniz 5th paper, § 99, Alexander, 1956, pp. 87–88)

With this interpretation of the inelastic collision, Leibniz answers an objection that contradicted his general law of conservation of living force (Appendix B). In contrast, Clarke not only argues that motion disappears in the case at stake, but he also denies any law of conservation. This radical opposition to a general law of conservation appears clearly in the discussion between both Leibniz and Clarke about natural theology.

[61] See also Thomson and Tait (1867, §§ 263, 268, 269), Lodge (1879, § 1; 1885, p. 482).

R. Lopes Coelho, *What Is Energy?*, History of Physics,
https://doi.org/10.1007/978-3-031-51855-3

At that time, it was necessary to show that a thesis about nature was in accordance with the theology. In this context, Leibniz argues that God is a perfect artificer. He created the world as a perfect clock, which works by itself since its creation. Clarke counter-argues: to admit that the world works for itself is to put God outside the government of the world. In that case, God would be like a nominal monarch. He would be king in name, but without influence in the kingdom. But there is more.

Clarke argues that there are no natural powers. A spring, for example, oscillates because God makes it oscillate. It does not oscillate by itself. The planets gravitate around the sun because God or an invisible created power performs these movements, still according to Clarke.[62] Now, if there are no natural powers,[63] then nature cannot function on its own.

Conclusion. In the Leibniz-Clarke controversy, we have two opposite theses concerning a conservation law. On one side, there is a law of conservation, that is, nature works by itself, independent of God (dependent at Creation, but not afterwards); and on the other, there is no law of conservation, as nature does not work by itself but depends completely on God. This latter view represents Newton's viewpoint. Thus, there is no reason to attribute the law of energy conservation to Newton.

References

Alexander, H. G. (Ed.). (1956). *The Leibniz-Clarke Correspondence*. Philosophical Library.

Koyré, A., Cohen, I. B. (1962). Newton and the Leibniz-Clarke Correspondence. *Archives Internationales d'Histoire des Sciences*, *15*, 63–126.

Lodge, O. (1879). An attempt at a systematic classification of the various forms of energy. *Philosophical Magazine*, *8*, 277–286.

Tait, P. G. (1865). Note on the History of Energy. *Philosophical Magazine*, *XXIX*, 55–57.

Thomson, W., Tait, P. (1867). *Treatise on Natural Philosophy* (Vol. I). Oxford University Press.

[62] Clarke's 3rd Reply, § 17, Clarke's 4th Reply, §§ 45–6.

[63] Clarke's 5th Reply, §§ 107–9.

Appendix M
On the Discovery of Energy

According to some physicists, we do not know what energy is (Sect. 1.1). Others define energy, but this definition has been criticized (Sect. 5.2). Since there is no clear concept of what it is, it becomes difficult to search the past for what concerns energy. What historians have done is to place in the history of energy what concerns the past of what they connect to the concept of energy. As the concept varies, depending on the author, it is not surprising that historiographical results about the energy concept differ from each other.[64] Such a divergence is illustrated below with a recurrent theme in the literature, the discovery of energy.

Mach defended the thesis that energy was an "almost simultaneous" discovery (Mach, 1896, p. 227). The period of discovery would have started with Carnot's manuscripts, which are prior to 1833. Mach explained the almost simultaneity through a set of trigger factors. These would have been the conviction of the impossibility of *perpetuum mobile* and the meaning of work (ibid. p. 226).

Mach's thesis has not been followed by other historians of energy.[65] The idea of the simultaneous discovery of energy reappears, however, in Kuhn's paper 'Energy conservation as an example of simultaneous discovery'. Mach spoke of 'almost simultaneous' because there was a time interval. Kuhn speaks of 'simultaneous discovery', but also presupposes a time interval. This goes from Carnot's manuscripts to 1854. Kuhn's paper was criticized for a variety of reasons.[66] He had, however, an appreciable repercussion.[67]

Kuhn presented 12 energy discoverers, which he systematizes into three groups of four authors:

---

[64] This divergence becomes specially clear if the history of the historiography of energy is considered (Mach 1872, Berthold 1875, Weyrauch 1885, etc. Chap. 1, Footnote 3).

[65] For example, Helm (1898), Haas (1909), Hiebert (1962) (see also Jammer 1963, pp. 164–6), Theobald (1966).

[66] Boyer (1959), Elkana (1970), Heimann (1973), Heimann (1974), Cantor (1975).

[67] Brush (1970), Rutherford et al. (1981), Breger (1982), Tetens (1987), Schirra (1989), Torreti (1999), Saslow (2020).

R. Lopes Coelho, *What Is Energy?*, History of Physics,
https://doi.org/10.1007/978-3-031-51855-3

1. Mayer, Joule Colding and Helmholtz (1842–47)
2. Sadi Carnot, Séguin, Holtzmann and Hirn (before 1833–54)
3. Mohr, Grove, Faraday and Liebig (1837–44). (Kuhn, 1959, p. 321).

The 1st group gathers the generality of formulation and concrete quantitative applications.[68] The 2nd group admits that heat and work are quantitatively interchangeable and computed an equivalent. The 3rd group sees the world of phenomena as manifesting a single force that could appear in many forms but never created or destroyed in all its transformations.

Mohr (group 3) is regarded as a discoverer of energy because he claimed that in addition to the 54 elements of chemistry, there was an agent in nature that was the force (Appendix G). How can one justify the inclusion of Mohr for this reason? If we admit that energy is an entity that acts in nature, it is understandable that Mohr appears in the history of energy. Now, according to Ostwald, energy is an agent. It is the cause of events (Sect. 4.3). Therefore, following Ostwald, we can admit energy as an entity. However, according to Feynman et al., (1963), Duit (1987), Hecht (2000), among others energy is not an entity. Following these authors, we have no reason to take Mohr as a discoverer for having argued that there is such a force. The historian's point of view is thus decisive in relation to the question of whether an author is a discoverer.

Let us continue with Holtzmann, who appears in group 2. He is considered a discoverer of energy because he calculated the unit of heat in terms of work. The value obtained coincides with Mayer's mechanical equivalent of heat (Appendix F). He did not say, however, that there was an entity that is an agent in nature, like Mohr. On the contrary, he admitted that the calculation of the mechanical value of the unit of heat was based on the expansion of heat. Therefore, Holtzmann did not discover a necessary characteristic of energy, which is the one for which Mohr is credited as a discoverer.

Thus, if Mohr is taken as a discoverer, the characteristic that he found for force must be a characteristic of energy. However, if this is a characteristic of energy, then one cannot attribute the discovery to Holtzmann, who found no agent. For the same reason, in order to give Holtzmann the merit of energy discoverer, it is necessary that the value of the mechanical equivalent of heat is a characteristic of energy. Nevertheless, if the calculation of the mechanical value of work is a necessary characteristic of energy, then we cannot attribute the discovery to Mohr, who did not calculate this value.

What is at issue here is the following. If one admits that energy has certain characteristics and states that A is a discoverer of energy, we expect that A had discovered that which has these characteristics. If we take A as a discoverer because he discovered only one of the characteristics, then we will have to say that he was not a discoverer of energy, but he discovered one of the characteristics of energy. Nevertheless, for this to have *historical meaning*, it is necessary that this characteristic was

---

[68] With regard 'quantitative applications' of this group, we have the following: Mayer and Joule determined the mechanical equivalent of heat; Colding did not calculate it; Helmholtz used it to make predictions.

subsequently used by someone else to discover the others. In the case at stake, it would be as follows: Mohr discovered the agent on the basis of which, for instance, the mechanical value of heat was arrived at. In such a situation, Mohr did not discover the mechanical value of heat, but is in a chain of events that led to that value. In that case, there were a reason to include him in the history of the discovery of energy. This was not, however, the case: neither Mayer nor Joule nor Holtzmann relied on Mohr's agent.

Of course, it could have happened that a third person had used Mohr's agent, Holtzmann's mechanical value of the unit of work and reached the conservation of energy. In this case, the two would be in a chain of events in the history of energy. However, this did not happen either. In sum, neither Mohr nor Holtzmann are in the chain of events that led to the conservation of energy. There is, therefore, no reason to take them as discoverers of energy.

Let us return briefly to the history of energy to continue the analysis. According to Mayer's axiom, the quantity of force in phenomena remains constant. Therefore, in a phenomenon of heat and work, there will have to be an equivalent between the two. (Otherwise, there were no conservation of force.) Joule admitted that heat was motion. Thus, he looked for a constant of proportionality between the two types of motion. In this search, he determined the mechanical value of heat (Joule 1843). Holtzmann calculated the work done by heat (as a substance) and determined the unit value of heat in terms of work. Let us add the values obtained in 1845: the mechanical equivalent of heat determined by Mayer was 367 kg m (3.60 J); by Joule, 798 lb ft (4.29 J); the value of the unit of heat in terms of work was 374 kg m (3.67 J).

So far, we have seen that different ways led to the calculation of a value, which is now called the 'mechanical equivalent of heat'. Since different means led to a value, we cannot say that all those who had determined the value by then, discovered energy. Let us look at this in detail.

Mayer's logical sequence can be given as follows: the conservation of force (his axiom) implies that there is a mechanical equivalent of heat. The search for this equivalent led to a given value. Therefore, the conservation of force led to the value.

$$
\begin{aligned}
&(Conservation\, of\; force => MEH_{concept})\; and\\
&(MEH_{concept} => MEH_{value}).\, Therefore,\\
&(Conservation\, of\; force => MEH_{value}).
\end{aligned}
$$

Since the premises are true, the conclusion holds, that is, (conservation of force $=> \mathrm{MEH}_{\mathrm{value}}$).

Let us now use this conclusion as a premise in the following syllogism:

1. conservation of force $=>$ $\mathrm{MEH}_{\mathrm{value}}$.
2. Now, Joule and Holtzmann arrived at this value. (This is true.)
3. Therefore, they discovered the conservation of force.

This reasoning is fallacious.

If one could only arrive at that value by conservation of force, then whoever arrived at the value had used conservation of force. Nevertheless, this is not the case. One arrived at the value from different premises, driven by different ideas.

This allows us to understand what happened in Kuhn's paper. Kuhn established the characteristics of energy. By discovering one of them, the author would enter the history of energy. The logic of this is as exposed earlier:

1. conservation of energy leads to the mechanical equivalent of heat;
2. some discovered the *value of* the equivalent;
3. then they discovered energy.

The reasoning is fallacious. (It is because of this fallacy that Holtzmann appears as the discoverer of energy, even though he admitted a principle (heat is substance) that was annihilated by the admission of energy.)

Like Mach, Kuhn also presents a set of factors to explain the simultaneity of the discovery. The trigger factors are, however, different. Kuhn's are:

- conversion processes,
- the concern with engines
- German philosophy of nature (*Naturphilosophie*).[69]

Several authors have criticized these factors. There is, however, a preliminary point to keep in mind. According to Kuhn, there was no simultaneous discovery. There was not because "no two of our men even said the same thing" (Kuhn, 1959, p. 322). "What we see in their works is not really the simultaneous discovery of energy conservation" (ibid. 323). So we have the following. The 3 factors serve to explain the simultaneity of the discovery. The discovery was not really simultaneous. Therefore, the factors are meant to explain what did not really happen. Therefore, Kuhn's factors have no historical relevance.

While Kuhn (1959), Brush (1970), Rutherford et al. (1981) spoke of 12 energy discoverers, other authors advocated lower numbers.[70] At the other end of the spectrum, there are theses with only one author. For some authors it was Mayer, for others, Joule and some argue that it was Helmholtz. This last thesis, which will be considered next, is from the twentieth century.

According to Elkana, "Helmholtz has proved correctly and generally the law of conservation of energy, at the time (1847)" (Elkana, 1974, p. 16). Some years later, Harman wrote: "Helmholtz's work on the conservation of energy was important not only in providing a mathematical formulation of the energy principle, but also [...]" (Harman, 1982, p. 44). In Torretti's *Philosophy of Physics*, Helmholtz was the first who clearly and precisely exposed energy as a fundamental physics principle (Torretti, 1999, p. 183). It is true that among the texts in Chap. 2, Helmholtz's paper is the one that makes the most use of mathematics. It is, however, important to see what he did with it.

[69] Brush (1970) added a fourth factor, the wave theory.

[70] Gross (1898), Helm (1898), Ostwald (1912 [1908]), Fox (1969), Lloyd 1970, Caneva (1993), Kragh (2009), Cahan (2012), Caneva (2021).

He expresses the conservation of force for the case of falling bodies,

$$\frac{1}{2}mv^2 = mgh \tag{M.1}$$

What follows is based on this. The first step of this generalization leads to the assumption of central forces (Sect. 2.4.1). In the case of electric and electromagnetic phenomena with current, it is not the previous equation but an equation of the form

$$\Delta living\ force = \Delta tension\ force$$

that he used. Indeed, the calculation of force, distance and velocity (Eq. M.1) in the case of phenomena that are not completely observable does not take place. Helmholtz looks at an electric circuit and takes one part as tension force and another part as living force. The advantage he points out in this procedure is heuristic. This leads to new results. Helmholtz's prediction was correct in some cases (Bevilacqua, 1993, p. 331). In conclusion, the ideas about Helmholtz's formulation of the conservation principle mentioned above are unfounded.[71]

The sense of discovery

When one speaks of 'discovery of energy' one presupposes that energy was hidden and is no longer.[72] If by discovery is meant bringing to light what was hidden in nature, then there is no discovery of energy, because, as we have seen, such an entity was never discovered (Chap. 6). What came from experiments and was new was the mechanical equivalent of heat.

The mechanical equivalent of heat not only served the thesis of the conservation of energy, but also justified the principle of equivalence, which did not use the concept of energy (Sect. 5.1.1). Therefore, the interpretation of the experiments through energy is not the only possible one. Thus, the question of the discovery of energy is ultimately the discovery of a given way of interpreting phenomena.

References

Berthold, G. (1875). Notizen zur Geschichte des Principes der Erhaltung der Kraft. *Monatsbericht der Königlich-Preussischen Akademie der Wissenschaten zu Berlin*, 577–586.

Bevilacqua, F. (1993). Helmholtz' Ueber die Erhaltung der Kraft. In D. Cahan (Ed.), *Hermann von Helmholtz and the foundations of the Nineteenth-century Science* (pp. 291–333). University of California Press.

[71] See also Bevilacqua (1993, pp. 321, 324).

[72] Some authors have this idea, for example, Kuhn: "Energy *is* conserved; nature behaves that way." (Kuhn, 1959, p. 323).

Boyer, C. B. (1959). Commentary on the Papers of Thomas S. Kuhn and I. Bernard Cohen. In M. Clagget (Ed.), *Critical Problems in the History of Science* (pp. 384–390). Wisconsin University Press.
Breger, H. (1982). *Die Natur als arbeitende Maschine: Zur Entstehung des Energiebegriffs in der Physik 1840–1850*. Campus Verlag.
Brush, S. (1970). The Wave Theory of Heat: A Forgotten Stage in the Transition from the Caloric Theory to Thermodynamics. *British Journal for the History of Science*, *5*, 145–167.
Cahan, D. (2012). The Awarding of the Copley Medal and the 'Discovery' of the Law of Conservation of Energy: Joule, Mayer and Helmholtz revisited. *Notes Rec. R. Soc.*, *66*, 125–139.
Caneva, K. L. (1993). *Robert Mayer and the conservation of energy*. Princeton University Press.
Caneva, K. L. (2021). *Helmholtz and the Conservation of Energy: Contexts of Creation and Reception*. MIT Press.
Cantor, G. (1975). William Robert Grove, the correlation of forces, and the conservation of energy. *Centaurus*, *19*, 723–290.
Duit, R. (1987). Should energy be illustrated as something quasi-material? *International Journal of Science Education*, *9*, 139–145.
Elkana, Y. (1970). The conservation of energy. *Arch. Int. d'Histoire des Sciences*, *90–91*, 30–60.
Elkana, Y. (1974). *Discovery of the Conservation of Energy*. Hutchinson.
Feynman, R. P., Leighton, R. B. & Sands, M. (1963). *The Feynman Lectures on Physics*. Addison-Wesley.
Fox, R. (1969). James Prescott Joule (1818–1889). In J. North (Ed.), *Mid-nineteenth-century Scientists* (pp. 72–103). Pergamon Press.
Gross, T. (1898). *Robert Mayer und Hermann v. Helmholtz*. M. Krayn.
Haas, A. (1909). *Die Entwicklungsgeschichte des Satzes von der Erhaltung der Kraft*. Hölder.
Harman, P. M. (1982). *Energy, Force, and Matter*. Cambridge University Press.
Hecht, E. (2000). *Physics: Calculus* (2 ed., Vol. 1). Brooks/Cole.
Heimann, H. (1973). Conversion of Forces and the conservation of Energy. *Centaurus*, *18*, 147–161.
Heimann, H. (1974). Helmholtz and Kant: the metaphysical Foundations of Ueber die Erhaltung der Kraft. *Studies in History and Philosophy of Science*, *5*, 205–238.
Helm, G. (1898). *Die Energetik nach der geschichtlichen Entwicklung*. Veit & C.
Hiebert, E. N. (1962). *Historical Roots of the Principle of Conservation of Energy*. University of Wisconsin Department of History.
Jammer, M. (1963). The Factorisation of energy. *British Journal for the Philosophy of Science*, *14*, 160–166.
Kragh, H. (2009). Conservation and controversy: Ludvig Colding and the imperishability of "forces". *RePoSS: Research Publications on Science Studies*, *4*, 1–27.
Kuhn, T. S. (1959). Energy conservation as an example of simultaneous discovery. In M. Clagget (Ed.), *Critical Problems in the History of Science* (pp. 321–356). Wisconsin University Press.
Lloyd, J. T. (1970). Background to the Joule-Mayer Controversy. *Notes and Records RSL*, *25*, 211–225.
Mach, E. (1872). *Die Geschichte und die Wurzel des Satzes von der Erhaltung der Arbeit*. Calve.
Mach, E. (1896). *Principien der Wärmelehre. Historisch-kritisch entwickelt*. J. A. Barth.
Ostwald, W. (1912). *Die Energie* (2 ed.). J. A. Barth.
Rutherford, F. J., Holton G., Watson, F. G. (1981). *Project Physics: Text*. Holt, Rinehart and Winston.
Saslow, W. M. (2020). A History of Thermodynamics: The Missing Manual. *Entropy*, *22*(1), Article 77.
Schirra, N. (1989). *Entwicklung des Energiebegriffs und seines Erhaltunskonzepts*. [Justus-Liebig-Universität]. Giessen.
Theobald, D. (1966). *The concept of energy*. Spon.
Tetens, H. (1987). *Experimentelle Erfahrung: Eine wissenschaftstheoretische Studie über die Rolle des Experiments in der Begriffs- und Theoriebildung der Physik*. Felix Meiner Verl.
Torretti, R. (1999). *The Philosophy of Physics*. Cambridge University Press.
Weyrauch, J. (1885). *Das Princip der Erhaltung der Energie seit R. Mayer*. Teubner.

Appendix N
On Energy Teaching

We have seen that there is no clear idea of what energy is (Sect. 5.2). Therefore, it is understandable that it is difficult to explain what it is. If it is difficult to expose the concept, we cannot expect students to understand it. This sequence of propositions reflects what research in science education has taught us.

For decades, explanations of energy in high-school and university textbooks have been criticized by science teaching experts,[73] which denotes the difficulty in explaining the concept. Research literature on student misconceptions and reasons for the misunderstandings is ample, which accounts for the students' difficulties.[74] Teaching methods have been developed in order to avoid misconceptions.[75] Nevertheless, if there is a problem with the concept to be taught, pedagogy can hardly overcome it.

Let us move on to the content of what has been taught. The explicit definition of energy that appears in textbooks tells us that energy is the capacity of doing work (Sect. 5.2.3). This definition has been criticized.[76] Beyond this definition, there is an implicit definition of energy, which is the one that emerges from the conservation principle: energy is something that can neither be destroyed nor created, but only transformed (Sect. 5.2.1). Duit (1987) pointed out, however, some inconveniences of the concept that takes energy as something quasi material. According to Beynon, there is so much confusion with energy "because it is not treated as an abstract physical quantity but something real, just like a piece of cheese" (Beynon, 1990, p. 315). Empirical educational research shows alternative ideas, such as 'Energy is

---

[73] Lehrman (1973), Sexl (1981), Duit (1981), Hicks (1983), Duit (1987), Bauman (1992), Chrisholm (1992), Arons (1999), Cotignola et al. (2002), Doménech et al. (2007), Lancor (2014), Hecht (2019), Bächtold (2021), Schulz and Kalman (2023).

[74] Watts (1983), Duit (1986), Nicholls and Ogborn (1993), Cotignola et al. (2002), Berg (2008), Jewett (2008a), Neumann and Nordine (2023).

[75] Solomon (1985), Trumper (1990, 1991, 1997), Prideaux (1995), Jewett (2008b), Besson and Ambrosis (2014), Aguiar et al. (2018), Meli et al. (2022).

[76] Lehrman (1973), Sexl (1981), Duit (1981), Hicks (1983), Kemp (1984), Doménech et al. (2007), Hecht (2019).

R. Lopes Coelho, *What Is Energy?*, History of Physics,
https://doi.org/10.1007/978-3-031-51855-3

fuel' or 'Energy is stored within objects' (Nicholls & Ogborn, 1993, p. 73; Prideaux, 1995, p. 278). Doménech et al. deconstruct ideas which could lead to an interpretation of energy as something possessed by the objects themselves (Doménech et al., 2007, pp. 51–53). All criticized topics concern the concept of energy as a real thing.

Some physics teaching experts prefer Feynman's concept: energy is not a concrete thing and energy conservation is a mathematical principle.[77] According to Feynman, the law of conservation of energy

> "states that there is a certain quantity, which we call energy, that does not change in the manifold changes which nature undergoes. That is a most abstract idea, because it is a mathematical principle; it says that there is a numerical quantity which does not change when something happens". (Feynman et al., 1963, Sect. 4.1)

This is in accordance with Mayer's axiom: the magnitude force, now 'energy', remains constant in physical and chemical phenomena. Mayer's axiom is based on the mechanical equivalent of heat, which comes from experiments (Sect. 2.1.2). Therefore, it is not necessary to resort to Feynman's idea to approach the energy concept. We can, however, understand the use of 'mathematical principle'. Mayer's original idea, which I have just used, is gone. The contemporary principle of energy conservation justifies conservation through an entity, which can neither be destroyed nor created (Sect. 2.1.2.1) Now, Feynman argued that such an entity does not exist. Moreover, he taught that it is not known what energy is. Thus, there remained the mathematical part of the principle, the conservation of quantity. According to the present study, it is not necessary to change the status of the principle—from a physical to a mathematical principle—to talk about the conservation of energy (Chap. 6).

Duit (1987, p. 145) pointed out that Feynman's concept is too abstract for teaching at basic level. It is understandable that this approach is difficult for those starting out, just as the concept of energy from Mayer's axiom can also be difficult for them. There is, however, a way out of this problem that consists of using the phylogenetic development for ontogenetic purposes. To introduce energy, we can use Mayer's first experiment, the agitation of water (Sect. 2.1.1). He did not give precise indications about the set up used, but it is possible to recreate it. A test tube containing 10 ml of water that was isolated and shaken for three minutes, showed a temperature increase of about half a degree Celsius. What Mayer read from the experiment was: motion transformed into heat. This reflects the common observation: I stirred the water, and it got warmer. The student will certainly understand what is meant by 'transformation'.

So far, we have had a qualitative approach. If you want to move on to a quantitative one (probably at a later stage), you can use Joule's paddle wheel experiment. When we show this experiment, the student already knows that stirring water gives heat. Now in Joule's experiment, the water is agitated, as the glass calorimeter experiment clearly shows. Therefore, there must be heat. What we need to know is how much heat was obtained by the falling bodies in Joule's experiment. For this, we will have to go on to measurements.

If we finally measure the work done and the heat produced, then we can put them into an equation based on an idea like 'cause = effect'. If we want to move on to

---

[77] Duit (1987), Prideaux (1995), Arons (1999).

energy, we will have to tell the students that we call two different things energy: weights at a certain height and the heat developed.[78]

One might object that by this I am not saying what energy is, but what I call energy. In fact, this is exactly what Mayer said could be done.[79] The reader might rightly think that textbooks also teach what is called energy. It is shown there that this is potential energy and that is kinetic energy, etc. There is, however, a difference: textbooks tell us what is called energy, but they do not tell us that energy is an assigned name, a homogenization term (Chap. 6). Without this information, it is left open what energy is. (It is, therefore, understandable that at the end of some explanations, students ask, 'but what is energy anyway?'[80]) If they know, however, that 'energy' is an assigned name and how that term was assigned, the doubt disappears. What needed to be said has been said. Of course, the student could, however, ask: 'if the name is given, then could it have been any other?' The answer is yes. Not only could it have had another name, but it also had another name at the beginning. If the student still asks, 'would it be possible not to name it at all?' the answer is also positive. The example comes to us from the authors who used the equivalence principle instead of the energy conservation principle.

In our century, there has been a widespread attempt to use energy as if it were a substance, even though the authors know that it is not. Brewe writes:

> "Treating energy with a conceptual metaphor of a substancelike quantity that can be stored and transferred provides students and instructors conceptual resources that contribute to the development of useful energy conceptions.
>
> However, simply including this metaphor for energy is not sufficient to promote energy as a viable way of modeling physical systems. The curriculum needs to be reorganized and refocused on energy as a central, coherent theme [...] Examples of student reasoning using energy to address common conceptions and to solve problems captures the power of the reasoning tools available with implementation of the energy-as-substance framework". (Brewe, 2011, p. 13)

The metaphor is a literary figure. It insinuates but does not define. If we want scientific language to be precise, we cannot use metaphors. The reason for the use of the energy-as-substance is the following: with this metaphor the student does what he is supposed to do. We can now achieve this same goal without the metaphor. The student does what he is intended to do because he understands what it is about, since this can be clearly exposed. (The teacher is expecting more indications and empirical work with students, but all this is beyond the scope of this book.)

---

[78] The steps in this sequence will naturally have to be adapted by the teacher.

[79] "was insbesondere die Kräftefrage anbelangt, so handelt es sich ja zunächst nicht darum, was ein "Kraft" für ein Ding ist, sonder darum, welches Ding wir "Kraft" *nennen* wollen." (Mayer, 1851, p. 35).

[80] See Hecht (2019, p. 498).

References

Aguiar Jr, O., Sevian, H., El-Hani, C. N. (2018). Teaching About Energy: Application of the Conceptual Profile Theory to Overcome the Encapsulation of School Science Knowledge. *Science & Education 27*, 863–893.

Arons, A. B. (1999). Development of energy concepts in introductory physics courses. *Am. J. Phys., 67*, 1063–1067.

Bächtold, M. (2021). Introducing Joule's Paddle Wheel Experiment in the Teaching of Energy: Why and How? *Foundations of Science, 26*, 791–805.

Bauman, R. P. (1992). Physics that textbook writers usually get wrong. *Phys Teacher, 30*, 264–269.

de Berg, K. (2008). The Concepts of Heat and Temperature: The Problem of Determining the Content for the Construction of an Historical Case Study which is Sensitive to Nature of Science Issues and Teaching–Learning Issues. *Sci & Educ, 17*, 75–114

Besson, U., Ambrosis, A. (2014). Teaching Energy Concepts by Working on Themes of Cultural and Environmental Value. *Sci & Educ, 23*, 1309–1338.

Beynon, J. (1990). Some myths surrounding energy. *Phys Educ, 25*, 314–316.

Brewe, E. (2011). Energy as a substance like quantity that flows: Theoretical considerations and pedagogical consequences. *Physical Review Special Topics—Physics Education Research, 7(020106)*, 1–14.

Chrisholm, D. (1992). Some energetic thoughts. *Phys Educ, 27*, 215–220.

Cotignola, M. I., Bordogna, C., Punte, G., Cappannini, O. M. . (2002). Difficulties in Learning Thermodynamic Concepts: Are They Linked to the Historical Development of This Field? *Sci & Educ, 11*, 279–291.

Doménech, J. L., Gil-Pérez, D., Gras-Marti, A., Guisasola, J., Martínez-Torregrosa, J., Salinas, J., Trumper, R., Valdés, P., Vilches, A. (2007). Teaching of energy issues: a debate proposal for a global reorientation. *Sci & Educ, 16*, 43–64.

Duit, R. (1981). Understanding energy as a conserved quantity – remarks on the article by R. U. Sexl. *Eur J Sci Educ, 3*, 291–294.

Duit, R. (1986). *Der Energiebegriff im Physikunterricht*. IPN, Abt. Didaktik d. Physik.

Duit, R. (1987). Should energy be illustrated as something quasi-material? *International Journal of Science Education, 9*, 139–145.

Feynman, R. P., Leighton, R. B. & Sands, M. (1963). *The Feynman Lectures on Physics* (Vol. 1). Reading: Addison-Wesley.

Hecht, E. (2019). Understanding energy as a subtle concept: A model for teaching and learning energy. *Am. J. Phys., 87*, 495–503.

Hicks, N. (1983). Energy is the Capacity to do Work – or is it? *Phys Teacher, 21*, 529–530.

Jewett, J. W. J. (2008a). Energy and the Confused Student III: Language. *The Physics Teacher, 46*, 149–153.

Jewett, J. W. J. (2008b). Energy and the Confused Student IV: A global approach to energy. *The Physics Teacher, 46*, 210–217.

Kemp, H. R. (1984). The concept of energy without heat and work. *Phys Educ, 19*, 234–240.

Lancor, R. (2014). Using Metaphor Theory to Examine Conceptions of Energy in Biology, Chemistry, and Physics. *Sci & Educ, 23*, 1245–1267.

Lehrman, R. (1973). Energy is not the ability to do work. *Am J Phys, 60*, 356–365.

Mayer, J. R. (1851). *Bemerkungen über das mechanische Aequivalent der Wärme*. Landherr.

Meli, K., Koliopoulos, D., Lavidas, K. (2022). A Model-Based Constructivist Approach for Bridging Qualitative and Quantitative Aspects in Teaching and Learning the First Law of Thermodynamics. *Sci & Edu 31*, 451–485.

Neumann, K., Nordine, J. C. (2023). Energy (In M. F. Taşar and P. R. L. Heron, (Eds.), *The International Handbook of Physics Education Research: Learning Physics*, Vol. 4–1). AIP Publishing.

Nicholls, G., Ogborn, J. (1993). Dimensions of children's conceptions of energy. *Int J Sci Educ, 15*, 73–81.
Prideaux, N. (1995). Different approaches to the teaching of the energy concept. *School Sci Rev 77*, 49–57.
Trumper, R. (1990). Being constructive: an alternative approach to the teaching of the energy concept – part one. *Int J Sci Educ, 12*, 343–354.
Trumper, R. (1991). Being constructive: an alternative approach to the teaching of the energy concept – part two. *Int J Sci Educ, 13*, 1–10.
Trumper, R. (1997). Applying conceptual conflict strategies in the learning of the energy concept. *Res Sci Tech Educ, 15*, 5–18.
Schulz, R. M., Kalman, C. S. . (2023). Philosophy of physics: Its significance for teaching and learning. In M. F. Taşar and P. R. L. Heron (Eds.), *The International Handbook of Physics Education Research: Special Topics* (pp. 12-11–12-30). AIP Publishing.
Sexl, R. U. (1981). Some observations concerning the teaching of the energy concept. *Eur J Sci Educ, 3*, 285–289.
Solomon, J. (1985). Teaching the conservation of energy. *Phys Educ, 20*, 165–170.
Watts, D. M. (1983). Watts, D. M. (1983) Some alternative views of energy. Phys Educ 18, 213–217. *Phys Educ 18*, 213–217.

SPRINGER NATURE

GPSR Compliance

The European Union's (EU) General Product Safety Regulation (GPSR) is a set of rules that requires consumer products to be safe and our obligations to ensure this.

If you have any concerns about our products, you can contact us on ProductSafety@springernature.com

In case Publisher is established outside the EU, the EU authorized representative is:

Springer Nature Customer Service Center GmbH
Europaplatz 3
69115 Heidelberg, Germany

The manufacturer's authorised representative in the EU is Springer Nature Customer Service Centre GmbH, Europaplatz 3, 69115 Heidelberg, Germany. If you have any concerns regarding our products, please contact ProductSafety@springernature.com

Printed and bound by CPI Group (UK) Ltd, Croydon, CR0 4YY
01/10/2026
02225898-0014